BRIEF TRIGONOMETRY

A TEXT IN TWENTY ASSIGNMENTS

ARTHUR R. CRATHORNE

GERALD E. MOORE

UNIVERSITY OF ILLINOIS

PRINTED IN THE UNITED STATES OF AMERICA

ISBN: 978-0-9851721-3-8

Digitally Reproduced in 2012 by
CONVERPAGE Digital Reproductions
23 Acorn Street
Scituate, MA 02066
www.converpage.com

PREFACE

This text book aims to give the essentials of plane trigonometry and logarithms in twenty reasonable assignments. The course in trigonometry is not shortened by leaving certain vital topics to the student but by abbreviating or omitting topics which some instructors prefer to give when occasion demands in courses in more advanced mathematics. These topics are inverse trigonometric functions and trigonometric equations, which are discussed briefly, and complex numbers and hyperbolic functions, which are omitted entirely.

In the chapters on logarithms and the solution of triangles, four place tables only are used and emphasis placed on interpolation.

The exercises and problems under each assignment are so planned that either the odd or the even numbered exercises may be assigned, the odd numbers having the answers given, and the even numbers without answers. An answer book giving answers to the even numbered exercises and problems is available.

For those teachers who prefer a longer course, a set of supplementary exercises and problems for each assignment and a general miscellaneous set of problems are given following the body of the text.

The miscellaneous problems and the index given at the end of the book are each designed for purposes of a thorough review, especially if the student follows the instructions given at the beginning of the index.

A. R. C.
G. E. M.

CONTENTS

	Page
Introduction	1
Assignment	
1. Angles	3
2. Trigonometric Functions	7
3. Special and Complementary Angles	10
4. Fundamental Identities	13
5. Right Triangles	18
6. Extension of Definitions	23
7. Unit Circle and Line Values	28
8. Graphs and Equations	33
9. Other Identities	38
10. Double and Half Angles	42
11. Product Formulas	46
12. Logarithms	48
13. Logarithms, Use of Tables	51
14. Logarithms (Applications)	55
15. Solution of Triangles Using Logarithms	57
16. Solution of Triangles	60
17. Solution of Triangles (Case II)	63
18. Solution of Triangles (Case III)	67
19. Solution of Triangles (Areas)	71
20. Solution of Triangles (Case IV)	74
Supplementary Exercises and Problems	77
Answers	97
Tables	105
Index	119

INTRODUCTION

> **trigonometry:** That branch of mathematics treating of the relations holding among the sides and angles of triangles.

This is the substance of most dictionary definitions of Trigonometry, but the usefulness of Trigonometry is not confined to the solution of triangles alone. The theoretical aspect of Trigonometry is not only indispensable in many branches of mathematics but also plays an important role in physics, astronomy, navigation (in air or on the sea), and electrical theory of alternating currents.

One of the most common applications is in surveying. Other problems, such as the simple pendulum, the motion of a projectile, simple harmonic motion, the theory of light and sound, the analysis of wave theory, the drawing of certain kinds of maps, and structural engineering, also make much use of the class of functions we are going to study.

The purpose of this brief text is to introduce the student to the fundamental notions and concepts of Trigonometry, to illustrate some of its uses, and to familiarize the student with the basic aspects through the solution of a variety of problems.

The student should recall certain facts from plane geometry, namely:

1. A ratio is an indicated quotient and is usually written $\frac{a}{b}$.

2. A proportion is a statement of equality between two ratios. Thus, $\frac{a}{b} = \frac{c}{d}$.

3. The Pythagorean Theorem: The square on the hypotenuse of a right triangle is equal to the sum of the squares on the other two sides.

4. If two triangles are similar, the ratios of corresponding sides are equal.

5. The length of the circumference of a circle is equal to π times the diameter:
$$C = \pi d = 2\pi r.$$

6. The area of a triangle equals one-half the product of the base by the altitude.

The student should provide for himself three simple instruments: a protractor, compasses and a ruler.

ASSIGNMENT 1

Angles

1. Angles.

We shall assume that the student is familiar with the concept of angles as presented in elementary geometry. That is, we may think of an angle as a figure formed by two intersecting lines. Again, we may think of an angle as being formed by two line segments radiating in different directions from a common point. Both of these aspects may be brought together in the following definitions which will enable us to extend our concept of angles, as is necessary in trigonometry.

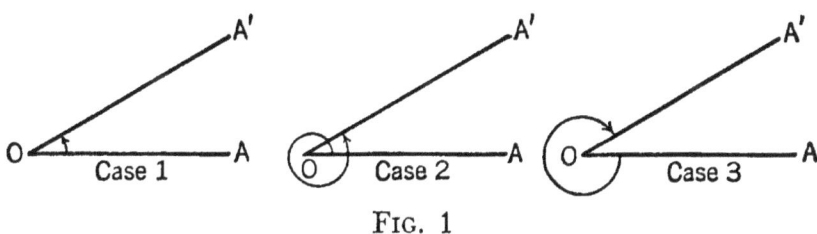

FIG. 1

DEFINITION: Suppose that a line segment OA rotates from its initial position to some new position OA' in a **counterclockwise** direction. Such a rotation is said to **generate** a **positive** angle with an **initial** side OA and a **terminal** side OA'. A **clockwise** rotation generates a **negative** angle. (See Fig. 1.)

The student should realize that an angle thus defined is in keeping with his previous conception of angles, and at the same time an extension or generalization of the concept of angles. If the line OA' is perpendicular to the line OA, we say that we have generated a right angle or 90°. If $A'OA$ forms a straight line, we say that we have generated a straight angle or 180°.

Furthermore, in Fig. 1, case 1, suppose that the angle AOA' is 30°. Also suppose that in each case the initial and terminal sides have the same positions. Then case 2 represents an angle of 390° and case 3, an angle of − 330°.

Exercises

Construct the following angles:

1. 45°	**4.** − 150°	**7.** 420°
2. 135°	**5.** − 90°	**8.** 225°
3. − 225°	**6.** 270°	**9.** − 390°

2. Units of angular measurement.

There are two common systems of angular measurement:

(a) The sexagesimal system (degrees, minutes and seconds),
(b) Radian measure.

The **sexagesimal** system is based on the division of a complete circumference into 360 divisions, so that a central angle of one fourth of a revolution contains 90°, etc. Each degree is divided into 60 minutes, so that 1° = 60′. Each minute is divided into 60 seconds, so that 1′ = 60″. The student is already familiar with this system since it was studied and used in plane geometry.

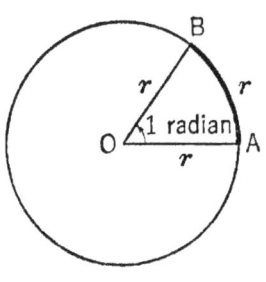

FIG. 2

The **radian** measure of an angle is based on the following considerations.

One starts with a circle of radius r. If the length of the radius is curved so that it may be placed in the position AB along the circumference (see Fig. 2), then the angle subtended at the center of the circle is called **one radian.**

This raises the question as to how many radians equals 360°. We know from geometry that the circumference C of a circle of radius r is given by

$$C = 2\pi r.$$

UNITS OF ANGULAR MEASUREMENT

This relationship may be written $\frac{C}{r} = 2\pi$, which says that the radius may be laid off as an arc 2π times, starting from a point such as A. Consequently,

$$2\pi \text{ radians} = 360°, \text{ or}$$
$$\pi \text{ radians} = 180°. \tag{1}$$

$$1 \text{ radian} = \frac{180°}{\pi} = \frac{180}{3.1416} = 57.2956° +. \tag{2}$$

From equation (1) we also have

$$180° = \pi \text{ radians}.$$

$$\therefore 1° = \frac{\pi}{180} = \frac{3.1416}{180} = .0175 \text{ radian}. \tag{3}$$

From equations (2) and (3) we have relationships which enable us to change radians to degrees, and degrees to radians.

As a result of the concept of the radian measure of an angle, it follows that we have an important ratio which always gives the radian value; namely,

$$\text{angle} = \frac{\text{arc}}{\text{radius}}.$$

If we designate the angle by θ, the arc by s, and the radius by r, this becomes

$$\theta = \frac{s}{r}. \tag{4}$$

EXAMPLE 1. Find the angle subtended by an arc of 14 inches on a circle of radius 3 inches.

SOLUTION: From equation (4), $\theta = \frac{14}{3} = 4\frac{2}{3}$ radians.

EXAMPLE 2. A wheel 7 inches in diameter makes 100 revolutions per minute. How far does a point on its circumference travel in 5 minutes?

SOLUTION: The radius of the wheel is $3\frac{1}{2}$ in. The point will travel $2\pi \cdot 3\frac{1}{2} = 7\pi$ in. in each revolution. In one minute it will travel 700π inches, and in 5 minutes, $3500\pi = 3500(\frac{22}{7}) = 11,000$ inches (approx.).

Exercises

1. Express each of the following angles in radian measure:
   30°, 45°, 60°, 90°, 120°, 135°, 150°, 180°, 210°, 270°, 300°, 315°, 330°, 360°.

2. Express each of the following angles in degrees:
   $$\frac{\pi}{7}, \frac{3\pi}{4}, \frac{\pi}{8}, -\frac{\pi}{3}, 6\pi, -2\pi, 1.6\pi, .3\pi.$$

3. A railroad curve is laid out along a circle of 1200 feet radius. If the curve subtends an angle of 139°, find the length of the curve.

4. The pendulum of a grandfather's clock is 32 inches long. It swings through an arc of $9\frac{1}{2}$ inches. Find the angle through which it swings, in radians.

5. An airplane propeller is 8 feet long from tip to tip and makes 2000 revolutions per minute. What is the linear velocity of a point on the tip of one blade? (That is, at what rate is the tip of the blade traveling in feet per minute?)

6. An automobile wheel (including the tire) is 32 inches in diameter. If the wheel makes 10 revolutions per minute, how fast is the car traveling in miles per hour?

7. Find the radius of a circle if an arc of $\frac{\pi}{4}$ inches subtends an angle of $22\frac{1}{2}°$.

8. State rules for changing degrees to radians, and radians to degrees.

─────────── ASSIGNMENT 2 ───────────

Trigonometric Functions

3. The trigonometric functions of an acute angle.

We begin by giving a restricted definition of the trigonometric functions which applies only to acute angles. We shall consider a **right** triangle with angles A, B, and C, and sides a, b, c opposite these respective angles, the right angle being C.

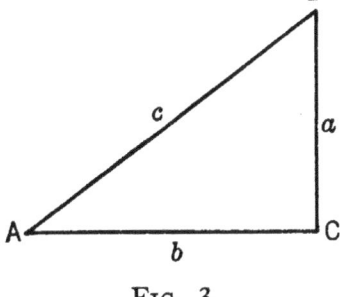

Fig. 3

Six ratios can be formed from these sides, and these **six ratios** are called the **trigonometric functions**. Their names are usually abbreviated as now shown for angle A.

$$\sin A = \frac{a}{c}. \quad \text{(Read ``sine of angle } A.\text{'')}$$

$$\cos A = \frac{b}{c}. \quad \text{(Read ``cosine of angle } A.\text{'')}$$

$$\tan A = \frac{a}{b}. \quad \text{(Read ``tangent of angle } A.\text{'')}$$

$$\cot A = \frac{b}{a}. \quad \text{(Read ``cotangent of angle } A.\text{'')}$$

$$\sec A = \frac{c}{b}. \quad \text{(Read ``secant of angle } A.\text{'')}$$

$$\csc A = \frac{c}{a}. \quad \text{(Read ``cosecant of angle } A.\text{'')}$$

These ratios can also be written in terms of the opposite side, adjacent side and hypotenuse when one confines the angle to an **acute** angle of a **right** triangle.

$$\sin A = \frac{a}{c} = \frac{\text{opposite side}}{\text{hypotenuse}}.$$

$$\cos A = \frac{b}{c} = \frac{\text{adjacent side}}{\text{hypotenuse}}.$$

$$\tan A = \frac{a}{b} = \frac{\text{opposite side}}{\text{adjacent side}}, \text{ etc.}$$

The student should **memorize** these ratios with their names, also write the ratios for angle B and draw right triangles in various positions. Practice, giving the ratios for each of the acute angles.

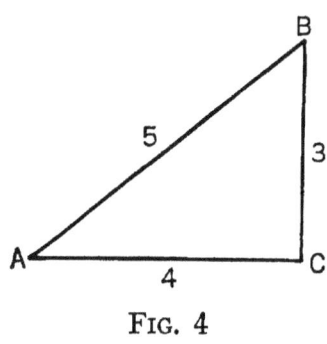

FIG. 4

EXAMPLE 1. Given a right triangle with $\sin A = \frac{3}{5}$. Write all the functions of angle A and also of angle B.

SOLUTION: Since $\frac{a}{c}$ is given as $\frac{3}{5}$, we will assume that $a = 3$ and $c = 5$. From $a^2 + b^2 = c^2$ we have $9 + b^2 = 25$ which yields $b^2 = 25 - 9 = 16$. Hence $b = 4$. Now from the definitions of the ratios we can write all the functions of angles A and B. Thus, $\tan A = \frac{3}{4}$, $\sec B = \frac{5}{3}$. The student should complete the list.

EXAMPLE 2. Given a right triangle with $\sin A = \frac{3}{5}$ and $c = 20$. Write all the functions of A and B.

SOLUTION: This differs from the preceding example in that we are given data which determine not only the ratios but also the size of the sides of the triangle. Thus we have $\frac{a}{c} = \frac{3}{5}$, but we are not free to choose $a = 3$, $c = 5$ as before. In fact we have $\frac{a}{c} = \frac{a}{20} = \frac{3}{5}$. Hence $a = 12$. Using $a = 12$, $c = 20$ gives $b = 16$. Now each ratio may be written as before.

Exercises

1. Compare the ratios obtained in Examples 1 and 2 above.
2. Write each of the functions for both acute angles of a right triangle, if given:

FUNCTIONS OF AN ANGLE

(a) $c = 9, b = 5$.
(b) $a = 3, b = 7$.
(c) $a = 7, b = 9$.
(d) $c = 11, a = 8$.
(e) $c = 16, b = 6$.
(f) $c = 8, a = 5$.

3. Which of the trigonometric functions are never greater than 1? Which are never less than 1?

4. Are there any functions which may be sometimes less than 1, sometimes greater than 1?

5. Discuss the values of sin A, cos A, and tan A, if the angle A is very nearly zero; nearly 90°.

6. Given cos $A = \frac{1}{2}$ and $b = 6$, construct the triangle and give the values of the functions of A.

7. Given tan $B = 3$, construct angle B.

---ASSIGNMENT 3---

Special and Complementary Angles

4. Functions of 45°, 30°, and 60°.

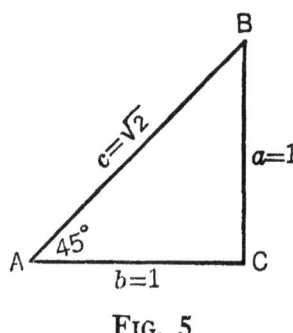

FIG. 5

To determine the functions of 45°, draw a **right isosceles** triangle and let $a = b = 1$. Then $c = \sqrt{2}$.

Hence,
$\sin 45° = \dfrac{1}{\sqrt{2}}$ $\csc 45° = \sqrt{2}$

$\cos 45° = \dfrac{1}{\sqrt{2}}$ $\sec 45° = \sqrt{2}$

$\tan 45° = 1$ $\cot 45° = 1$.

To determine the functions of 30° and 60°, draw an **equilateral** triangle ABC. Draw the perpendicular BD, thereby bisecting the angle B as well as bisecting AC. Call $AB = 2$, then $AD = 1$ and $BD = \sqrt{3}$. Hence,

$\sin 30° = \dfrac{1}{2}$ $\csc 30° = 2$

$\cos 30° = \dfrac{\sqrt{3}}{2}$ $\sec 30° = \dfrac{2}{\sqrt{3}}$

$\tan 30° = \dfrac{1}{\sqrt{3}}$ $\cot 30° = \sqrt{3}$.

We also have

$\sin 60° = \dfrac{\sqrt{3}}{2}$ $\csc 60° = \dfrac{2}{\sqrt{3}}$

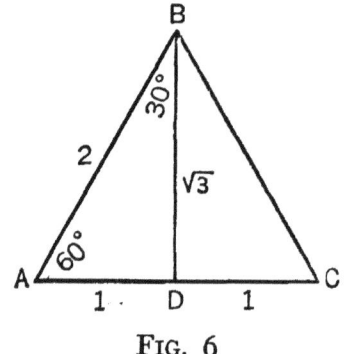

FIG. 6

COMPLEMENTARY RELATIONSHIPS

$$\cos 60° = \frac{1}{2} \qquad \sec 60° = 2$$

$$\tan 60° = \sqrt{3} \qquad \cot 60° = \frac{1}{\sqrt{3}}.$$

The values of the functions as given above are **exact**. **Approximate** values to four decimals may be obtained by using $\sqrt{2} = 1.4142$ and $\sqrt{3} = 1.7321$. However, one usually uses the radical form in these cases unless directed to do otherwise.

Exercises

1. Given an equilateral triangle whose sides are of length a. From the figure give the values of each of the functions of 60° and 30°.
2. If a right isosceles triangle has its two equal sides of length k, determine the functions of 45°.

By actual substitution, verify the following:

3. $\tan 60° = \dfrac{2 \sin 30°}{1 - \tan^2 30°}.$
4. $\sin 45° + \sin 60° = (\sin 30° + \cos 30°)^2.$
5. If $A = 30°$, verify $\sin 2A = 2 \sin A \cos A$.
6. If $A = 60°$, verify $\sin \dfrac{A}{2} = \sqrt{\dfrac{1 - \cos A}{2}}.$

5. Complementary relationships.

Let us arrange the six functions by pairs as follows:

$$\sin A, \cos A; \quad \tan A, \cot A; \quad \sec A, \csc A.$$

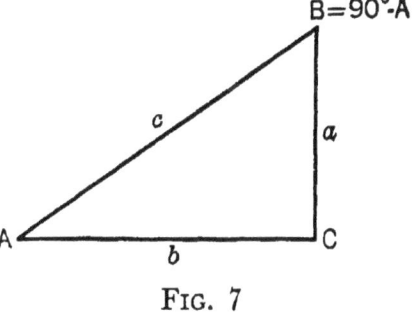

Fig. 7

Either function in a pair is called the **co-function** of the other member of the pair. The naming is more obvious if the student will write them, sine A, co-sine A; etc. The sine is the co-function of the cosine; the cosine is the co-function of the sine; and similarly for the other pairs.

Let us examine the situation from a triangle. The student should note that

$$\sin A = \cos B = \frac{a}{c}.$$

12 SPECIAL AND COMPLEMENTARY ANGLES

But $A + B = 90°$ and hence are complementary angles. The student should now examine a sufficient number of cases to be convinced that

co-functions of complementary angles are equal.

Thus, $\sin 30° = \cos 60°$. $\sin 25° = \cos(90° - 25°) = \cos 65°$.

Also, if $\tan A = \frac{3}{7}$, then the cotangent of the complement of A is also $\frac{3}{7}$, so that $\cot B = \frac{3}{7}$.

Exercises

1. Express each of the following as functions of its complementary angle:

$\sin 63°$, $\cos 38°$, $\tan 71°$, $\cot 25°$, $\sec 15°$, $\csc 81°$.

2. Find the acute angle x for which $\sin x = \cos(60° - x)$.

SOLUTION: $\sin x = \cos(90° - x)$, consequently

$$\cos(90° - x) = \cos(60° - x).$$

Since the cosines of the angles are equal, and the angles are acute, the angles are equal, and we have

$$90° - x = 60° - x, \quad \text{or} \quad x = 15°.$$

3. Find the acute angle for which $\cos 2x = \sin x$.

4. Find the acute angle for which $\tan 3x = \cot\left(x + \dfrac{\pi}{4}\right)$.

5. Find the angle A if $\sin(A + 45°) = \cos(A + 30°)$.

ASSIGNMENT 4

Fundamental Identities

6. Reciprocal and quotient relationships.

The reader is probably familiar with the concept of a number and its reciprocal. The reciprocal of 2 is $\frac{1}{2}$; the reciprocal of $\frac{1}{2}$ is $\frac{1}{\frac{1}{2}} = 2$. The reciprocal of a is $\frac{1}{a}$.

In general, two numbers are said to be **reciprocals** of each other when their product is unity.

Reciprocal relationships exist between certain pairs of the trigonometric functions. Thus,

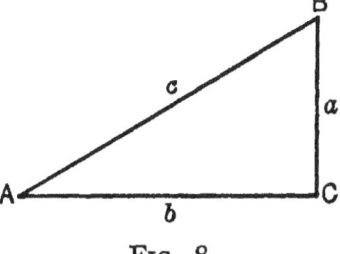

FIG. 8

$$\sin A \cdot \csc A = \frac{a}{c} \cdot \frac{c}{a} = 1,$$

and we may write

$$\sin A = \frac{1}{\csc A}, \quad \text{or} \quad \csc A = \frac{1}{\sin A}.$$

The student should verify each of the following reciprocal pairs, and then **memorize** them.

$$\text{I} \begin{cases} \sin A = \dfrac{1}{\csc A}. \\ \cos A = \dfrac{1}{\sec A}. \\ \tan A = \dfrac{1}{\cot A}. \end{cases} \qquad \text{II} \begin{cases} \csc A = \dfrac{1}{\sin A}. \\ \sec A = \dfrac{1}{\cos A}. \\ \cot A = \dfrac{1}{\tan A}. \end{cases}$$

FUNDAMENTAL IDENTITIES

Two other simple relationships connect the functions:

$$\text{III} \quad \tan A = \frac{\sin A}{\cos A} \quad \text{and} \quad \text{IV} \quad \cot A = \frac{\cos A}{\sin A}.$$

Relationship III may be verified as follows:

$$\frac{\sin A}{\cos A} = \frac{\frac{a}{c}}{\frac{b}{c}} = \frac{a}{c} \cdot \frac{c}{b} = \frac{a}{b}.$$

But by definition,

$$\tan A = \frac{a}{b}.$$

Hence

$$\tan A = \frac{\sin A}{\cos A}.$$

The student should verify relationship (IV) first, by using the definitions; second, by use of the reciprocal relationship for $\tan A$ and $\cot A$.

7. Squares of the functions.

There are three other important relationships connecting the squares of certain pairs of the trigonometric functions. These are:

$$\text{V} \quad \sin^2 A + \cos^2 A = 1.$$
$$\text{VI} \quad 1 + \tan^2 A = \sec^2 A.$$
$$\text{VII} \quad 1 + \cot^2 A = \csc^2 A.$$

To establish these relationships we use a right triangle with sides a and b and hypotenuse c. From the Pythagorean theorem, $a^2 + b^2 = c^2$.

By dividing each term of this equation by c^2,

$$\frac{a^2}{c^2} + \frac{b^2}{c^2} = 1 \quad \text{or} \quad \left(\frac{a}{c}\right)^2 + \left(\frac{b}{c}\right)^2 = 1.$$

Since by definition $\frac{a}{c} = \sin A$ and $\frac{b}{c} = \cos A$, we have

$$* \sin^2 A + \cos^2 A = 1.$$

---

* It is customary to write $(\sin A)^2$ as $\sin^2 A$ instead of $\sin A^2$. This last symbol is faulty in that it would seem to indicate the square of an angle, whereas $\sin^2 A$ indicates the square of the value of the function of the angle A.

SQUARES OF THE FUNCTIONS

Similarly, by dividing each term of $a^2 + b^2 = c^2$ by b^2 and a^2, respectively, one obtains the two equations:

$$1 + \left(\frac{a}{b}\right)^2 = \left(\frac{c}{b}\right)^2 \quad \text{or} \quad 1 + \tan^2 A = \sec^2 A,$$

$$1 + \left(\frac{b}{a}\right)^2 = \left(\frac{c}{a}\right)^2 \quad \text{or} \quad 1 + \cot^2 A = \csc^2 A.$$

These relationships have been shown to be true for acute angles of a right triangle. Although not shown here, these relationships are also true for any angle whatsoever.* That is, these relationships are true for every angle, and for that reason they are called the **fundamental identities.**

We now proceed to show how these identities are used to change a trigonometric expression of a given form into an equivalent expression of a different form. In many instances this sort of procedure results in a simplification of the form of the expression. The use of these identities in other courses, such as analytic geometry and calculus, is an important aspect of their usefulness in theoretical and applied mathematics. For this reason the student should memorize the identities (I) to (VII) and learn well how to use them.

EXAMPLE 1. Show that $\dfrac{1 + \sin x}{\cos x} = \sec x + \tan x$.

SOLUTION: Starting with the left-hand member one has, by writing as separate fractions,

$$\frac{1 + \sin x}{\cos x} = \frac{1}{\cos x} + \frac{\sin x}{\cos x}.$$

By identities (II), Art. 6, $\dfrac{1}{\cos x} = \sec x$. By (III), Art. 6, $\dfrac{\sin x}{\cos x} = \tan x$. Therefore, we may write

$$\frac{1}{\cos x} + \frac{\sin x}{\cos x} = \sec x + \tan x,$$

thus establishing the given relationship.

EXAMPLE 2. Show that $\sec x - \tan x \sin x = \cos x$.

SOLUTION: Expressing the left-hand member in terms of $\sin x$ and $\cos x$ gives

$$\sec x - \tan x \sin x = \frac{1}{\cos x} - \frac{\sin x}{\cos x} \cdot \sin x = \frac{1 - \sin^2 x}{\cos x}.$$

* See *Trigonometry*, by Crathorne & Lytle, Rev. Ed., Chap. 5, pp. 52–55.

16 FUNDAMENTAL IDENTITIES

From (V), Art. 7, $\quad 1 - \sin^2 x = \cos^2 x$

Therefore, $\quad \dfrac{1 - \sin^2 x}{\cos x} = \dfrac{\cos^2 x}{\cos x} = \cos x.$

EXAMPLE 3. Show that $\cos^4 x - \sin^4 x = 2\cos^2 x - 1$.

SOLUTION: Factoring the left-hand member as one would factor the algebraic expression, $a^4 - b^4 = (a^2 - b^2)(a^2 + b^2)$, gives

$$(\cos^2 x - \sin^2 x)(\cos^2 x + \sin^2 x) = \cos^4 x - \sin^4 x.$$

From (V), Art. 7, $\quad \sin^2 x = 1 - \cos^2 x$ and $\sin^2 x + \cos^2 x = 1$.

Hence, $\quad [\cos^2 x - \sin^2 x][\cos^2 x + \sin^2 x] = [\cos^2 x - (1 - \cos^2 x)][1]$
$$= \cos^2 x - 1 + \cos^2 x$$
$$= 2\cos^2 x - 1.$$

8. General directions for establishing identities.

1. Reduce one side of the expression by means of the fundamental identities until one obtains the required relationship. Or,
2. Reduce both sides of the proposed identity until each side has been reduced to a common expression. Or,
3. If no direct approach seems evident, reduce all expressions to sines and cosines and simplify both members by algebraic procedure. This should leave both members in the same form.

Method 1, above, is recommended in most instances, as it develops more skill and ability in the use of identities.

Exercises

Establish each of the following identities.

1. $\sin x \sec x = \tan x.$
2. $(1 + \tan^2 x)\cos^2 x = 1.$
3. $\cot^2 x - \cos^2 x = \cot^2 x \cos^2 x.$
4. $(\csc^2 x - 1)\sin^2 x = 1.$
5. $\dfrac{\sin x}{\csc x} = 1 - \dfrac{\cos x}{\sec x}.$
6. $\cos x \sqrt{\sec^2 x - 1} = \sin x.$
7. $\dfrac{\sin x}{1 + \cos x} + \dfrac{1 + \cos x}{\sin x} = 2\csc x.$

Hint: Reduce the left member to a single fraction with a common denominator.

8. $\csc^2 A - \csc^2 A \cos^2 A = 1.$
9. $\dfrac{\csc x + 1}{\cot x} = \dfrac{\cot x}{\csc x - 1}.$

Hint: Multiply both numerator and denominator of the left member by $\csc x - 1$, then proceed to use fundamental identities.

IDENTITIES

10. $\cot A + \dfrac{\sin A}{1 + \cos A} = \csc A.$
11. $\sin^2 x \sec^2 x = \sec^2 x - 1.$
12. $(1 - \sec^2 x)(1 - \csc^2 x) = 1.$
13. $\sec^4 x - \tan^4 x = \sec^2 x + \tan^2 x.$
14. $\sqrt{\dfrac{1 - \cos x}{1 + \cos x}} = \csc x - \cot x.$
15. $\tan A + \cot A = \sec A \csc A.$
16. $(\sin A + \cos A)^2 + (\sin A - \cos A)^2 = 2.$
17. $\cos^4 x - \sin^4 x + 1 = 2 \cos^2 x.$
18. $(\sec x + \tan x)^2 = \dfrac{1 + \sin x}{1 - \sin x}.$
19. $\sqrt{\dfrac{1 - \sin x}{1 + \cos x}} = \sec x - \tan x.$
20. $\cos A + \tan A \sin A = \sec A.$

──────ASSIGNMENT 5──────

Right Triangles

9. Use of trigonometric tables.

Up to the present time we have discovered how to determine the trigonometric functions of angles of 30°, 45°, 60°, etc. Tables of the values of the functions of angles for given degrees and minutes are in common use. Such a table begins on page 108.

Referring now to this table we see that

sin 33° = .5446, cos 28° = .8829, tan 45° = 1.0000, etc.

Many of the angles which we will use are not listed directly in the tables. For example, suppose we wish to find the value of sin 33° 24′. This angle does not appear, however 33° 20′ and 33° 30′ are given, and we shall assume that the value of sin 33° 24′ lies between the values which are given in the table. Thus:

$$10'\left\{ 4'\left\{ \begin{matrix} \sin 33° \ 20' = .5495 \\ \sin 33° \ 24' = \quad ? \\ \sin 33° \ 30' = .5519 \end{matrix} \right\} x \right\} .0024$$

We assume that the rate of change of the function is constant for a small angular change. Hence we may write the proportion $\frac{4'}{10'} = \frac{x}{.0024}$. On solving for x we obtain

$$x = \frac{4(.0024)}{10} = .00096 \text{ (which we shall call .0010)}.$$

Consequently, if the sin 33° 20′ = .5495 has .0010 added to it, we obtain sin 33° 24′ = .5505. This process is called **interpolation**, since it involves placing values between known values of the table.

18

SOLUTION OF RIGHT TRIANGLES

As a second illustration, suppose we wish to find the value of cos 21° 33′. This angle lies between the tabular values 21° 30′ and 21° 40′. Hence, as before, we have

$$10'\left\{3'\left\{\begin{array}{l}\cos 21° 30' = .9304\\ \cos 21° 33' = \quad ?\\ \cos 21° 40' = .9293\end{array}\right\}x\right\}.0011$$

This time we set up the proportion $\dfrac{3'}{10'} = \dfrac{x}{.0011}$, which yields $x = \dfrac{3(.0011)}{10} = .00033$ (which we shall call .0003).

Note that the cosine function has a **decreasing** * value as the angle increases. Consequently, the interpolation must be accomplished in this case by **subtracting** .0003 from the value of cos 21° 30′. This gives cos 21° 33′ = .9304 − .0003 = .9301.

Exercises

1. Find sin 37° 38′.
2. Find A, if cos A = .8778.
3. Find cos 43° 17′.
4. Find tan 72° 18′.
5. Find A, if sin A = .6528.
6. Find A, if tan A = .1721.

10. Solution of right triangles.

If we have given any two sides, or a side and one of the acute angles of a right triangle, we can always find the remaining dimensions of the triangle and its area. This problem of finding the remaining parts is called **solving the triangle.** Just how this is done can best be illustrated by examples.

EXAMPLE 1. Solve the right triangle, if given that
$$A = 37° \quad \text{and} \quad c = 38.2.$$
SOLUTION: $B = 90° - 37° = 53°$.

By definition sin $A = \dfrac{a}{c}$, so that

$$\sin 37° = \dfrac{a}{38.2}.$$

But sin 37° = .6018, so that
$$(38.2)(.6018) = 22.99 = a.$$

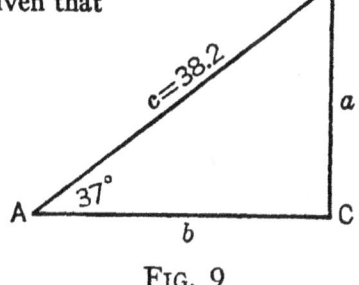

FIG. 9

---

* The sine and tangent functions increase for increasing values of the angle from 0° to 90°, whereas the cosine and cotangent decrease in the same interval as the angle increases.

20 RIGHT TRIANGLES

By definition, $\cos A = \dfrac{b}{c}$. Hence

$$\cos 37° = \dfrac{b}{38.2} \quad \text{or} \quad (38.2)(\cos 37°) = b$$

$$(38.2)(.7986) = 30.51 = b.$$

The area $= \dfrac{1}{2}ab = \dfrac{1}{2}(22.99)(30.51) = 350.76$.

Note that after the side a was found to equal 22.99, we might have used $a^2 + b^2 = c^2$ and solved for b. The student may easily verify that this would require much more labor than the method given.

Note also that in solving a triangle the part sought and the two given parts must involve one of the functions, hence the formula chosen should involve these three parts.

EXAMPLE 2. Given the sides $a = 28$, $c = 51.3$, to find the angle A and side b.

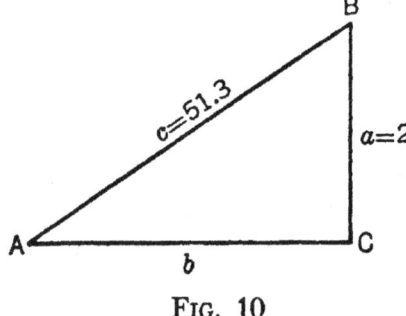

FIG. 10

SOLUTION: From the relation

$$\sin A = \dfrac{a}{c} = \dfrac{28}{51.3} = .5575$$

we find

* $A = \text{arc sin } .5575 = 33° 53'$.

To find b we may use either of the formulas

$$\dfrac{a}{b} = \tan A \quad \text{or} \quad \dfrac{b}{a} = \cot A.$$

To solve for b from the first of these involves a long division. Consequently, we shall use the second formula, which may be written

$$b = a \cot A = 28(1.4890) = 41.69$$

Exercises

In each of the following right triangles, two parts are given. Compute the remaining parts and the area.

1. $a = 17$, $b = 13$
2. $a = 29$, $c = 63$
3. $A = \dfrac{\pi}{3}$, $b = 24.3$
4. $c = 53$, $A = 62° 15'$

---

* Arc sin is called the inverse function notation, meaning that A is an angle whose sine is .5575. It is read, "A equals inverse sine .5575" or as written, "arc sin .5575."

SOLUTION OF RIGHT TRIANGLES

5. $B = 18° 5'$, $b = 27.2$

6. Show that if the area of a triangle is $\frac{1}{2}ab$, it can also be expressed as $\frac{1}{2}bc \sin A$.

7. Find the length of the side of a regular pentagon inscribed in a circle of radius 3 feet.

Hint: Draw a $\perp$ from the center to one of the sides, thus forming right triangles.

8. The angle of elevation of a chimney CB is $57°$, as observed from a point A which is 150 ft. from C and on a horizontal line with the base of the chimney. How high is the chimney? (**Angle of elevation** is defined as the angle between the line of sight and the horizontal line, if the point observed is **above** the horizontal.)

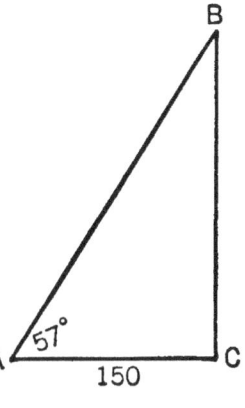

FIG. 11

9. A man M on an observation tower 110 ft. high notes that the angle of depression of an enemy outpost E is $2° 15'$. What is the distance,

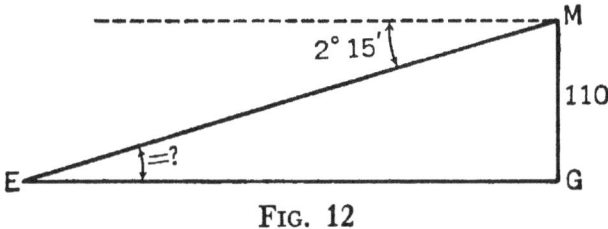

FIG. 12

measured along the ground, to the outpost? (**Angle of depression** is defined as the angle between the line of sight and the horizontal, if the point observed is **below** the horizontal.)

10. The sun casts a shadow 115 ft. long from a flagpole 175 ft. high. What is the angle of elevation of the sun? What is the approximate time of day if the sun rose at 6 A.M.?

11. An isosceles triangle (Fig. 13) has its vertical angle equal to $43° 20'$ and the base is 32.8 ft. Find the altitude, length of the equal sides, and one of the angles A or B.

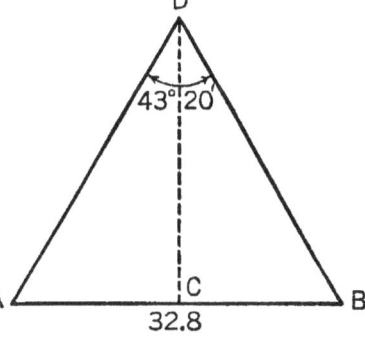

FIG. 13

12. A man observes that from where he stands, the angle of elevation of the top of a chimney is $47° 15'$. He walks backwards 25 ft. and finds that the angle of elevation is now $42° 25'$. How high is the chimney?

How far was the man from the chimney when he made the first observation?

Hint: $\dfrac{h}{d} = \tan 47° 15'$,

$\dfrac{h}{25 + d} = \tan 42° 25'$,

giving two equations which may be solved simultaneously for h and d.

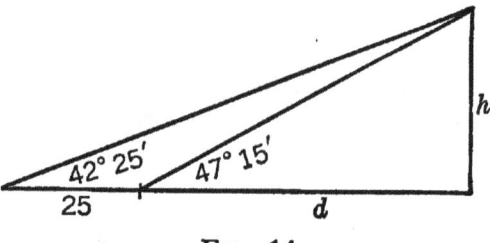

Fig. 14

13. The pilot of a seaplane wishes to determine the length of a small island. He notes that his altimeter reads 5000 feet just as he is directly above one end of the island. He estimates the angle of depression of the other end of the island to be 37° 15'. How long is the island?

14. Two observers on level ground and separated from each other by 2500 feet observe a plane at the same instant. The plane is above and just crossing the line between the observers. One of them finds the angle of elevation to be 62° 23'; the other finds the angle of elevation to be 41° 17' (Fig. 15). What is the altitude of the plane?

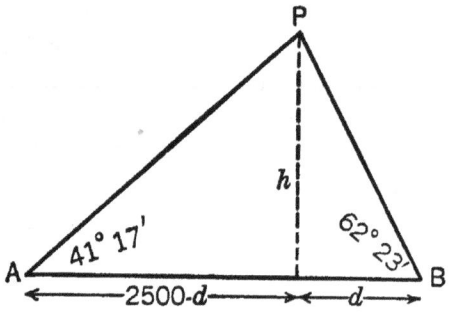

Fig. 15

15. A flagpole CD is mounted on the corner of a building (Fig. 16). An observer at O finds the angle of elevation of the top of the building to be 46° 15', the angle of elevation of the top of the pole to be 49° 16', and his distance from the building to be 243 feet. How high is the pole?

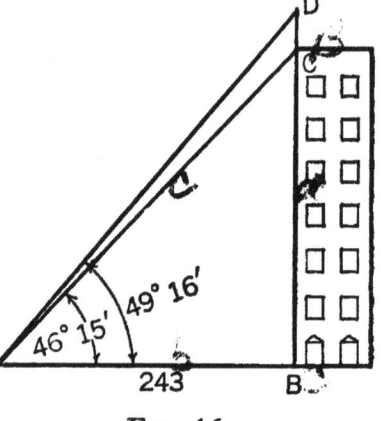

Fig. 16

---ASSIGNMENT 6---

Extension of Definitions

11. Rectangular coordinate system.

Points in a plane may be designated with reference to a pair of mutually perpendicular lines called the **X-axis** and **Y-axis**, and a suitably chosen unit length. The point O, in which the axes intersect, is called the **origin**. Every point in the plane can be named by an ordered pair of values (x, y) called the **coordinates** of the point. The x value is called the **abscissa**; the y value is called the **ordinate**. The distance of the point from the origin, such as OP, is called the **radius vector**. The x values are laid off as follows: **positive** to the right of O, **negative** to the left of O. The y values are laid off: **positive** above O, and **negative** below O. The plane is thus divided into four **quadrants,** as numbered in the order shown by Fig. 17.

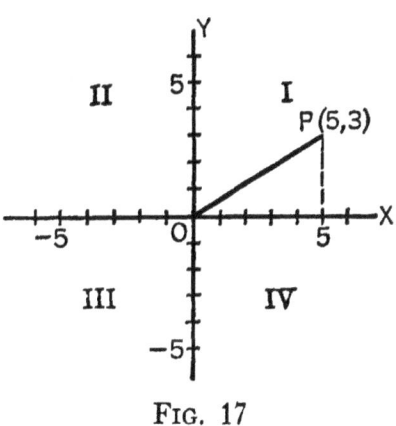

Fig. 17

12. Extension of the definitions of the functions.

If any point P in the plane is joined to O by a line, an angle is formed between the positive direction of the X-axis and the line OP.

We shall always measure angles by considering the vertex to be at the origin, the initial side along the positive X-axis, and

24 EXTENSION OF DEFINITIONS

the terminal line some line OP. Such an angle is said to be in **standard position** with reference to a rectangular coordinate system. We shall designate the angle by θ (the Greek letter "theta") and the length of the terminal side OP by r.

We have, by this scheme, another method of designating the coordinates of P, by giving the distance r and the angle θ. Such pairs of values (r, θ) are called the **polar coordinates** of a point P.

We have already seen that the values of the trigonometric functions are fixed if the angle is fixed, and therefore we may choose any point P on the terminal side of the angle, drop a perpendicular to the X-axis, and thereby form a triangle with sides x, y, and r. Now we can express the values of the trigonometric ratios in terms of these three quantities. The triangle thus formed is called the **triangle of reference** for the given angle (see Fig. 18).

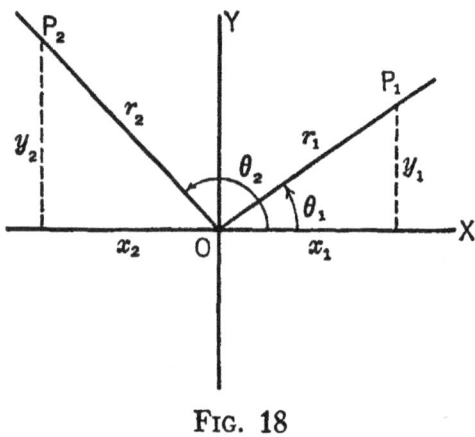

Fig. 18

With this set-up we may now give an extended definition of the trigonometric functions. Our new definition will include the acute angle definition (i.e., an angle in the first quadrant) but applies also to any angle whatsoever. That is, the angle may have its terminal side in any one of the four quadrants. The value of r will be considered **positive** in any quadrant when it is measured along the terminal side of the angle. (If r is measured along the backward extension of the terminal side of the angle, then r will be considered negative.)

The extended definitions of the functions follow, and should be **carefully memorized** by the student.

$$\sin \theta = \frac{y}{r} \qquad \csc \theta = \frac{r}{y}$$

$$\cos \theta = \frac{x}{r} \qquad \sec \theta = \frac{r}{x}$$

SIGNS OF THE FUNCTIONS

$$\tan \theta = \frac{y}{x} \qquad \cot \theta = \frac{x}{y}.$$

These definitions hold for all angles defined by a radius vector which does not coincide with one of the axes. Some cases, in which the terminal side of the angle lies on an axis, may involve division by zero,* and these cases will be considered later.

13. Signs of the functions.

The signs of the coordinates of a point are as follows in the four quadrants.

        1st quadrant: x is $+$, y is $+$
        2nd quadrant: x is $-$, y is $+$
        3rd quadrant: x is $-$, y is $-$
        4th quadrant: x is $+$, y is $-$.

For example, in Fig. 19,

$$\sin \theta_2 = \frac{y_2}{r_2} = \frac{+}{+} \text{ is positive,}$$

$$\cos \theta_2 = \frac{x_2}{r_2} = \frac{-}{+} \text{ is negative,}$$

$$\tan \theta_3 = \frac{y_3}{x_3} = \frac{-}{-} \text{ is positive, etc.}$$

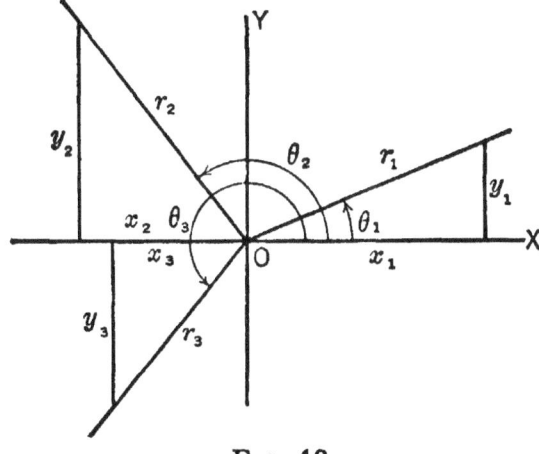

Fig. 19

---

* It is not possible to define division of a number by zero, but we can study what happens to a ratio when the denominator approaches zero.

The proper sign to be applied to each function in each quadrant can be assigned according to the scheme shown in Fig. 20.

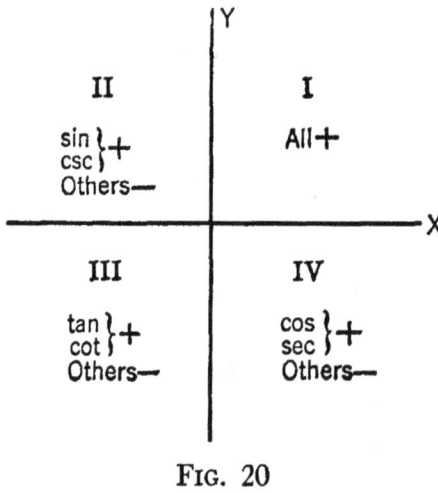

Fig. 20

Fig. 21

EXAMPLE 1. Give the values of sin 150°, cos 150°, tan 150°.

SOLUTION: Sketch the angle 150° and note that its ratios can be determined from our knowledge of functions of an angle of 30° (see Fig. 21). From the figure,

$$\sin 150° = \frac{1}{2}$$

$$\cos 150° = \frac{-\sqrt{3}}{2}$$

$$\tan 150° = \frac{1}{-\sqrt{3}}.$$

EXAMPLE 2. Given a point whose polar coordinates are $(\sqrt{2}, 225°)$, draw the angle thus determined and give the values of sin, cos, and tan.

SOLUTION: Sketch Fig. 22, and label the sides of the triangle of reference as shown.

Then,

$$\sin 225° = \frac{-1}{\sqrt{2}}$$

$$\cos 225° = \frac{-1}{\sqrt{2}}$$

$$\tan 225° = \frac{-1}{-1} = +1.$$

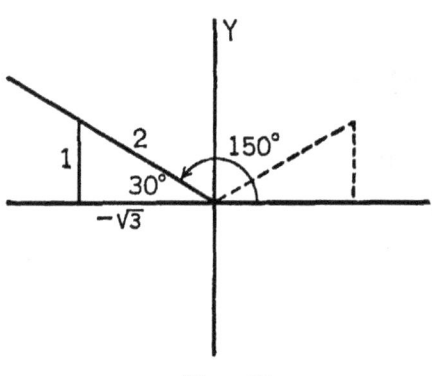

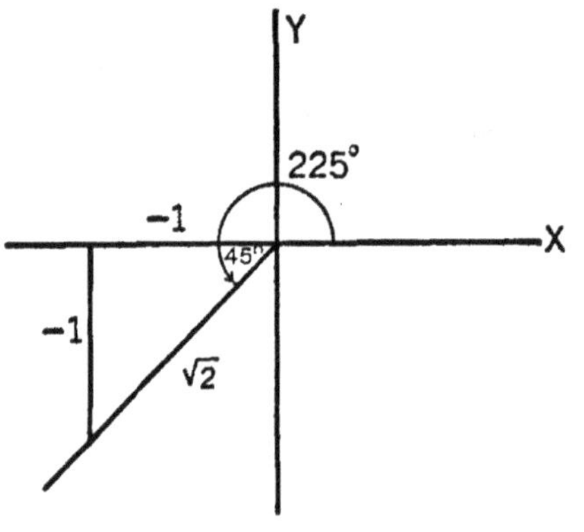

Fig. 22

Fig. 22 can also be used to determine the trigonometric functions of $-135°$, and they are the same as the functions of 225°.

SIGNS OF THE FUNCTIONS

Exercises

1. Join each of the following points to the origin, and give the values of the six functions for the positive angle thus formed.

   $P_1(3, 4)$, $P_2(7, 5)$, $P_3(-2, 6)$, $P_4(-3, -5)$, $P_5(4, -2)$

2. (a) A line joins the point $(5, 2)$ to the origin. Find the angle which it forms with the positive direction of the x-axis.
   (b) Do the same for each of the following points:
   $$(-5, 2),\ (-5, -2),\ (5, -2).$$
   Note that the angle found in part (a) plays an important role in each of the angles in part (b).

3. Using the lengths of the sides as were used in finding the functions of $30°$, $45°$, $60°$, and a properly chosen triangle of reference on a coordinate system, complete the following chart.

	30°	45°	60°	120°	135°	150°	210°	225°	240°	300°	315°	330°
sin												
cos												
tan												

4. Given $\sin A = \frac{3}{5}$ and that the angle A terminates in the second quadrant, find all the other functions of the angle A.

   Name the possible quadrants in which the angle A may terminate, if:

5. $\sin A = -\frac{3}{4}$.
6. $\cos A = \frac{2}{3}$.
7. $\tan A = \frac{5}{12}$, and $\sin A$ is positive.
8. What happens to the value of $\tan A$ if angle A approaches the value $0°$? If angle A approaches $90°$?
9. Show that the following relationships hold in whatever quadrant the angle θ terminates.

   $x = r \cos \theta$ $\qquad$ $x^2 + y^2 = r^2$

   $y = r \sin \theta$ $\qquad$ $\frac{y}{x} = \tan \theta$

10. Use these relationships of Exercise 9 to change the following coordinates from rectangular to polar coordinates (or vice versa).

    (a) $(3, 4)$ $\qquad$ (c) $(4, 45°)$

    (b) $(-1, \sqrt{3})$ $\qquad$ (d) $(3, 120°)$

―――――ASSIGNMENT 7―――――

Unit Circle and Line Values

14. Circle of radius one unit.

We have already seen on several occasions that the values of the trigonometric functions, being ratios, do not depend upon the lengths of the sides of the triangle of reference, but depend rather upon the angle. For this reason we may simplify much of trigonometry by considering a circle with center at the origin and radius equal to 1 unit. (See Fig. 23.)

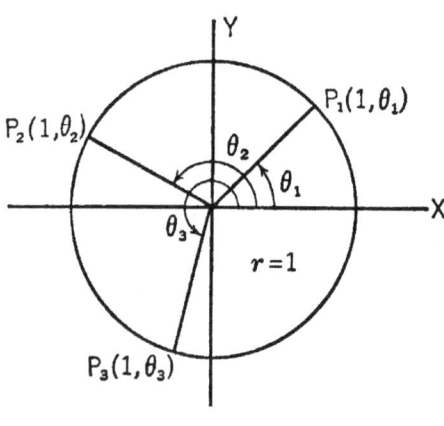

Fig. 23

Such a circle is called the **unit circle**. Suppose now that we lay off any angle θ. Then the polar coordinates of the point on the circle where the radius vector cuts the circle will always be $(1, \theta)$ regardless of the quadrant in which θ lies.

Thus from Fig. 23, the point P_1 has polar coordinates $(1, \theta_1)$, the point P_2 has coordinates $(1, \theta_2)$, etc. Therefore, if $\theta_1 = 45°$, the point P_1 has polar coordinates $(1, 45°)$ or what is the same thing in radian measure $\left(1, \dfrac{\pi}{4}\right)$.

15. Line values of the functions.

By making use of the unit circle we may easily find how the values of the trigonometric functions change with a change of

LINE VALUES OF THE FUNCTIONS

angle. This is accomplished by keeping one of the dimensions of the triangle of reference constantly equal to unity.

In Fig. 24, draw any angle θ whose terminal side cuts the unit circle in P and also cuts the tangent drawn to the circle at A in a point Q.

Consider the coordinates of P to be either $(1, \theta)$ or (x, y). We have then,

$$\sin \theta = \frac{y}{1} = y = MP$$

$$\cos \theta = \frac{x}{1} = x = OM$$

$$\tan \theta = \frac{y}{x} = \frac{AQ}{1} = AQ$$

$$\sec \theta = \frac{1}{x} = \frac{OQ}{1} = OQ.$$

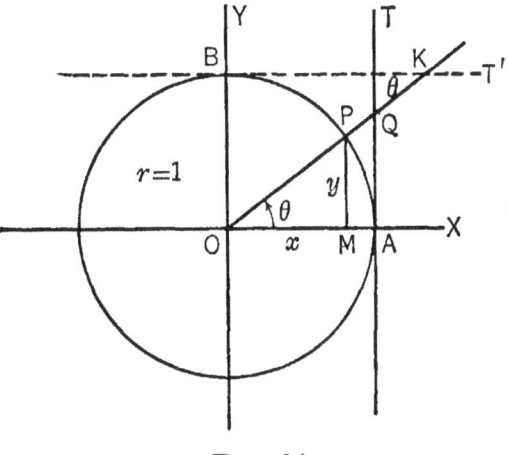

Fig. 24

By drawing a tangent at the complementary position B, as shown in Fig. 24, we have also

$$\cot \theta = \frac{BK}{OB} = \frac{BK}{1} = BK$$

$$\csc \theta = \frac{OK}{OB} = \frac{OK}{1} = OK.$$

If $\theta = 30°$, $PM = \sin 30° = \frac{1}{2}$.
If $\theta = 45°$, $AQ = \tan 45° = 1$, etc.

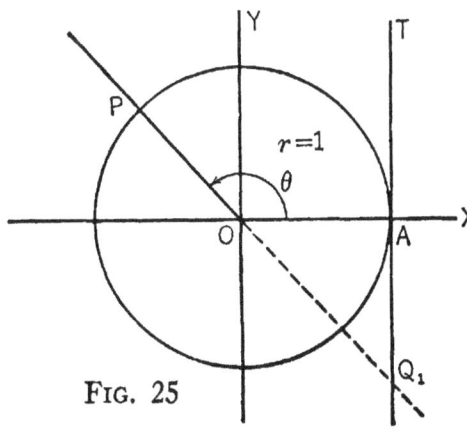

Fig. 25

If θ is an angle, say in the second quadrant, the terminal side of the angle will not cut the tangent line AT. However, the **backward projection** of the terminal side (dotted line, Fig. 25) will cut AT at Q_1. Hence the tangent of an angle in the second quadrant is negative, since AQ_1 is below the x-axis and measures the value of the tangent function. This procedure will be used for every case where the terminal side of the angle does not cut the lines AT or BT_1, and its use gives the proper sign to the value of the function, as already shown in Art. 13.

Thus, for a second quadrant angle,

$$\tan \theta = AQ_1 \text{ (and is negative)}$$
$$\sec \theta = OQ_1 \text{ (and is negative)}$$

For a third quadrant angle, the terminal side extended cuts AT in a positive length, but the secant is negative since it is measured on the extension of the terminal side of the angle.

Exercises

1. Draw a unit circle. Draw an angle in the second quadrant and show the line values of all six functions, determining whether positive or negative from the figure.
2. As in Exercise 1, using an angle θ in the third quadrant, also use an angle θ in the fourth quadrant.

16. Reduction formulas.

Tables of the values of the trigonometric functions give the values for angles ranging from 0° to 90°. We shall now show how these tables may be made use of to give the values of the functions for any angle whatsoever.

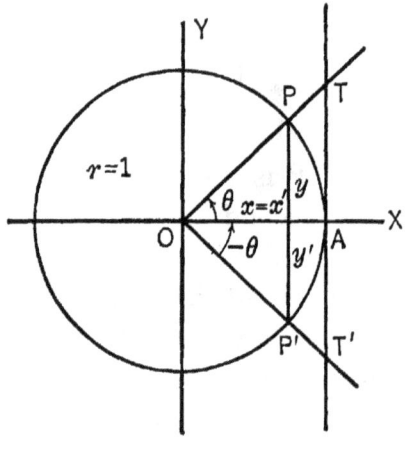

Fig. 26

For this purpose we shall use the line value concept of Art. 15.

Consider θ to be an acute angle, drawn in standard position in a unit circle (Fig. 26). Now construct an angle $(-\theta)$. Since the two reference triangles are congruent, we have, with respect to the coordinate system,

$$x = x', \; y = -y'$$

REDUCTION FORMULAS

From the line value point of view,

$$\sin(-\theta) = -\sin\theta \qquad \csc(-\theta) = -\csc\theta$$
$$\cos(-\theta) = \cos\theta \qquad \sec(-\theta) = \sec\theta$$
$$\tan(-\theta) = -\tan\theta \qquad \cot(-\theta) = -\cot\theta.$$

If we consider the angle $(-\theta)$ to be equivalent to $(360° - \theta)$, then we have, from this same figure,

$$\sin(360° - \theta) = -\sin\theta \qquad \csc(360° - \theta) = -\csc\theta$$
$$\cos(360° - \theta) = \cos\theta \qquad \sec(360° - \theta) = \sec\theta$$
$$\tan(360° - \theta) = -\tan\theta \qquad \cot(360° - \theta) = -\cot\theta.$$

Consider θ to be an acute angle and construct an angle $(180° - \theta)$.

From Fig. 27, and from consideration of the facts that $x = -x'$, $y = y'$,

$$\sin(180° - \theta) = \sin\theta$$
$$\cos(180° - \theta) = -\cos\theta$$
$$\tan(180° - \theta) = -\tan\theta,$$

and similarly for the other three functions.

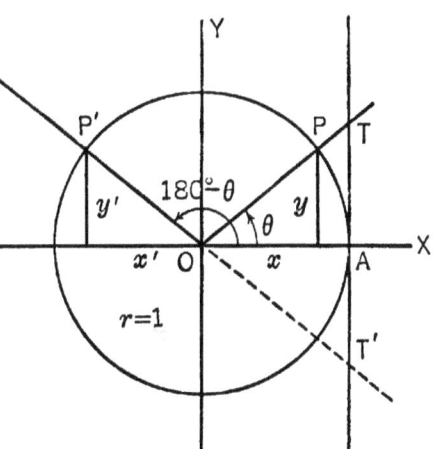

Fig. 27

Consider θ to be an acute angle and construct an angle $(180° + \theta)$, as in Fig. 28.

Now $x = -x'$, $y = -y'$ and AT coincides with AT'. Consequently,

$$\sin(180° + \theta) = -\sin\theta$$
$$\cos(180° + \theta) = -\cos\theta$$
$$\tan(180° + \theta) = \tan\theta,$$

and similarly for the other three functions.

The results of all these formulas can be summarized in a single statement given in this —

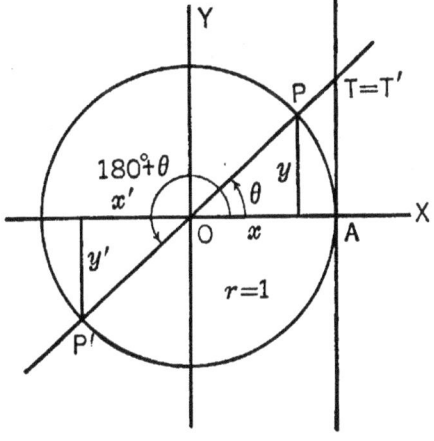

Fig. 28

32 UNIT CIRCLE AND LINE VALUES

RULE. Any function of any angle = the *same name* function of an acute angle θ, if one writes the given angle as $(n \cdot 180° \pm \theta)$, n being an integer. The **sign** of the result is determined from the **given angle** and the **given function** according to the rule for signs in Art. 13.

EXAMPLE 1. Give the value of sine, cosine and tangent of 210°.

SOLUTION: 210° may be written as 180° + 30°. In this case θ becomes 30°. Consequently,

$$\sin 210° = \sin(180° + 30°) = -\sin 30° = -\tfrac{1}{2}$$

(negative because sine is −, in 3rd quadrant)

$$\cos 210° = \cos(180° + 30°) = -\cos 30° = -\frac{\sqrt{3}}{2}$$

(negative because cosine is −, in 3rd quadrant)

$$\tan 210° = \tan(180° + 30°) = +\tan 30° = +\frac{1}{\sqrt{3}}$$

(positive because tangent is +, in 3rd quadrant)

EXAMPLE 2. Give equivalent values in terms of an acute angle, for sin 137°, cos 235°, cot 395°.

SOLUTION: $\sin 137° = \sin(180° - 43°) = +\sin 43°$
$\cos 235° = \cos(180° + 55°) = -\cos 55°$
$\cot 495° = \cot(360° + 35°) = +\cot 35°.$

Exercises

Give equivalent values for each of the following in terms of an acute angle θ.

1. sin 163°
2. cos 148°
3. cot 318°
4. tan 283°
5. sec 251°
6. cos 195°
7. csc 375°
8. sin 555°
9. tan 137°

10. sin (− 157°).   *Hint:* First write − 157° instead of + 203°.

11. Using the reduction formulas and rule, check the values in the chart, Art. 13, Exercise 3.

12. Show that if θ is an acute angle, the functions of $(90° + \theta)$ and $(270° \pm \theta)$ can be expressed in terms of the co-named functions of θ, with the proper sign.

ASSIGNMENT 8

Graphs and Equations

17. Graphs of sine, cosine and tangent functions.

The line values may also be used to obtain the graphs of the trigonometric functions. For this purpose the values of the functions will be drawn as ordinates corresponding to the values of the angles as abscissas.

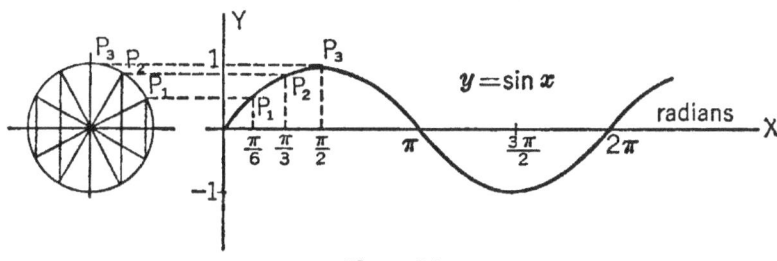

FIG. 29

As in Fig. 29, lay off angles of 0, $\frac{\pi}{6}$, $\frac{\pi}{3}$, etc., in a unit circle. Lay off on the x-axis the angles 0, $\frac{\pi}{6}$, $\frac{\pi}{3}$, etc., using $\pi = 3\frac{1}{7}$ units (approx.). Corresponding to each angular value, draw an ordinate equal to the line value of the function as obtained from the circle. Connect the ends of these ordinates by a smooth curve. This curve gives pictorially the shape of the graph of $y = \sin x$. Thus we see, as has already been shown, that the sine function is positive in the first two quadrants, negative in the third and fourth quadrants.

The graph of $y = \cos x$ may be obtained by a similar procedure, the only change necessary being to erect at each point of the x-axis an ordinate whose length is obtained from the line value of the cosine on the unit circle (see Fig. 30). The ordinates on the graph for P_1 and P_2 are OM_1 and OM_2, respectively.

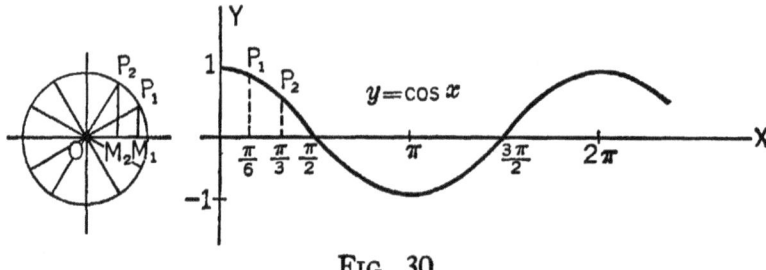

FIG. 30

The sine and cosine curves are of basic importance in wave theory, the study of harmonic motion, and alternating electric current theory.

Using the line values of the tangent function, one obtains a graph of $y = \tan x$, as shown in Fig. 31.

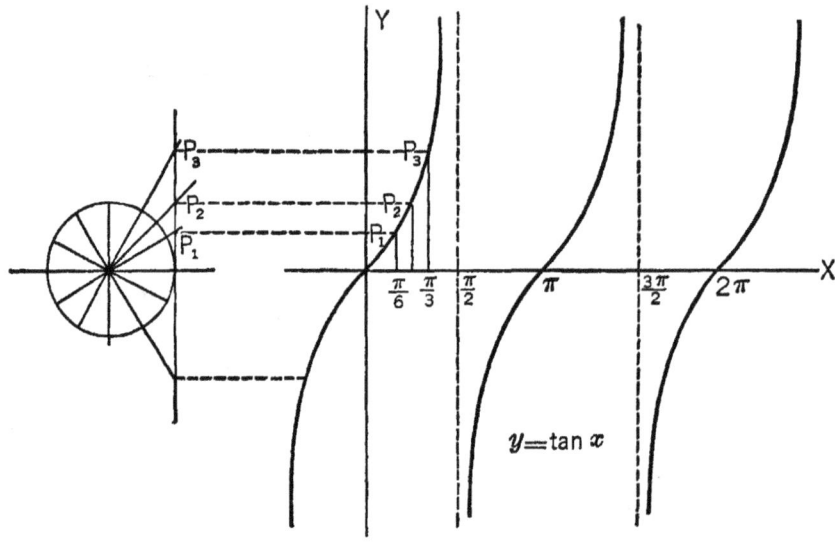

FIG. 31

The graph of $y = \tan x$ shows that as the angle increases toward 90°, the value of the tangent gets larger and larger. We see that the tangent becomes indefinitely large if the angle

VARIATION AND PERIODICITY

is acute and nearly 90°. Sometimes this is expressed by saying that tan 90° is infinite, usually written, tan 90° = ∞. The symbol "∞" is read "infinity," and the equality sign in this instance is not used in the ordinary sense of equality. Perhaps "=" in this instance should be read "becomes," since it involves a limiting process.

Exercises

1. Using the method outlined in this section, draw the graph of $y = \cot x$.
2. Draw the graph of $y = \sec x$.
3. Draw the graph of $y = \csc x$.
4. Show that the graphs of sin x, cos x and tan x may also be obtained from plotting the points of Art. 13, Exercise 3.

18. Values of the functions of quadrantal angles.

By means of line values, or from the graphs of the trigonometric functions, the following functional values are evident.

	0° or 360°	90°	180°	270°
sin	0	1	0	− 1
cos	1	0	− 1	0
tan	0	∞*	0	∞*
csc	∞*	1	∞*	− 1
sec	1	∞*	− 1	∞*
cot	∞*	0	∞*	0

19. Variation and periodicity of the functions.

A study of the graphs of the trigonometric functions shows that they repeat themselves in certain cycles. Such periodic behavior gives rise to a special terminology concerning such functions. They are said to be **periodic**.

The sine function repeats itself every interval of 2π radians (or every 360°). Its **period** is described by saying that it is of

---

* For the sense in which this symbol is used, see the last paragraph of Art. 17.

length 2π. The cosine function has also the period 2π. The tangent is also periodic, but its period is π.

One can easily trace the variation of the values of the functions by examining their graph. For instance, the sine function increases from value 0 at 0° to 1 at 90°, decreases from 1 at 90° to 0 at 180°, decreases from 0 at 180° to -1 at 270°, and then increases from -1 at 270° to 0 at 360°. If one considers the function for values of the angle from 360° to 720° (i.e., from 2π to 4π), the graph repeats itself in form for each such successive interval.

Similar remarks may be made concerning the periods of the cotangent, secant and cosecant. The student should verify from his solutions to problems 1, 2, and 3 of the Exercises, Art. 17, that:

> The period of the cosecant is 2π,
> The period of the secant is 2π,
> The period of the cotangent is π.

20. Direct vs. inverse functions.

We shall now call special attention to some important differences between direct and inverse functions.

Given sin 30°. There is but one value for this expression. It is a **ratio,** and its value is $\frac{1}{2}$.

Given $\theta = $ arc sin $\frac{1}{2}$. This is an inverse * function, and there are two **angles** between 0° and 360° which satisfy the equation, namely, 30° and 150°. In each revolution there are two angles coterminal with 30° and 150°, so that there are an unlimited number of angles for which the sine has the value $\frac{1}{2}$. Similarly, $\theta = $ arc tan (-1) has the values 135° and 315° in the first revolution.

These differences may be summarized by saying:

A direct function is single-valued and represents a ratio.

An inverse function is multiple-valued (always two solutions between 0° and 360°), and these values are angles.

* See the inverse function notation in Art. 10, p. 20.

21. Trigonometric equations.

An equation involving trigonometric functions for which we are required to find solutions is called a **trigonometric equation**. We shall now illustrate the way in which some simple types of these equations may be solved.

EXAMPLE 1. Find angles, less than 360°, which satisfy
$$\sin\left(x + \frac{\pi}{2}\right) = \frac{\sqrt{3}}{2}.$$

SOLUTION: $x + \frac{\pi}{2} = \text{arc sin } \frac{\sqrt{3}}{2}$. But, arc sin $\frac{\sqrt{3}}{2} = 60°$ or $120°$.

Therefore, $x + \frac{\pi}{2} = 60°$ or $x + \frac{\pi}{2} = 120°$.

So that, $x = 60° - \frac{\pi}{2} = 60° - 90° = -30°$, or $x = 120° - 90° = 30°$.

EXAMPLE 2. Find angles, less than 360°, which satisfy
$$2 \sin^2 \theta - \cos \theta - 1 = 0.$$

SOLUTION: One must first reduce this to an expression in terms of a single function. Since $\sin^2 \theta = 1 - \cos^2 \theta$,

$$2(1 - \cos^2 \theta) - \cos \theta - 1 = 0,$$
$$2 - 2 \cos^2 \theta - \cos \theta - 1 = 0,$$
$$-2 \cos^2 \theta - \cos \theta + 1 = 0,$$

and dividing by -1, $\quad 2 \cos^2 \theta + \cos \theta - 1 = 0.$

This can be solved by factoring as one would factor an algebraic expression into
$$(2 \cos \theta - 1)(\cos \theta + 1) = 0.$$

If $\;2 \cos \theta - 1 = 0$, $\cos \theta = \frac{1}{2}$ or $\theta = \text{arc cos } \frac{1}{2} = 60°$ or $300°$.
If $\;\cos \theta + 1 = 0$, $\cos \theta = -1$ or $\theta = \text{arc cos } -1 = 180°$.

Exercises

Solve for angles, less than 360°, which satisfy each of the following:

1. $\cos\left(x + \frac{\pi}{6}\right) = \frac{1}{2}.$
2. $\sin^2 x = \frac{1}{4}$
3. $5 \sin x = 3$
4. $\tan^2 x - 1 = 0$
5. $\sin x - \csc x = -\frac{3}{2}$
6. $\sin 2x = \frac{\sqrt{3}}{2}$
7. $2 \sin^2 x + 3 \cos x = 0$
8. $(2 \cos x - 1)(\sin x + 1) = 0$
9. $2 \tan^2 x + 3 \sec x = 0$
10. $\sin^2 x - \cos^2 x = \frac{1}{2}$

ASSIGNMENT 9

Other Identities

22. The sine and cosine of sums.

So far, we have been studying the functions of a single angle. In this article we begin the study of functions of angles which are the sum or difference of two angles.

More specifically, we shall now establish two important identities.

VIII. $\sin(x + y) = \sin x \cos y + \cos x \sin y.$
IX. $\cos(x + y) = \cos x \cos y - \sin x \sin y.$

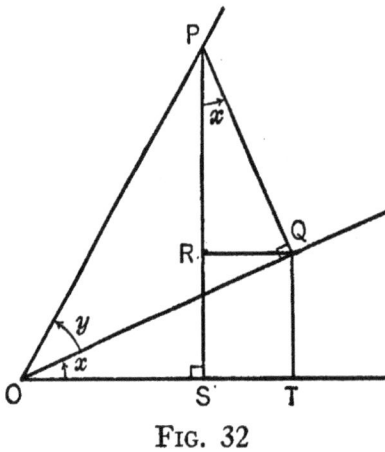

Fig. 32

We proceed to prove (VIII) for two **acute** angles x and y, whose sum $(x+y)$ is also **acute**.* In Fig. 32 let x and y be the two acute angles. From any point P on the terminal side of the sum of the angles, drop perpendiculars PQ and PS. Draw RQ, which is parallel and equal to ST. Draw QT parallel and equal to RS. Then $\triangle QPR = \triangle TOQ$.

By definition, $\sin(x+y) = \dfrac{SP}{OP}$, but we wish to derive other expressions for this ratio from the geometry of the figure.

* Although these identities (VIII) to (XXIII) are true for any angles whatever, the proofs given here are for acute angles. For the generalization to any angle, see *Trigonometry*, by Crathorne and Lytle, Rev. ed., pp. 58-70.

TANGENT OF SUM AND DIFFERENCE

$$\sin(x+y) = \frac{SP}{OP} = \frac{SR+RP}{OP} = \frac{TQ+RP}{OP} = \frac{TQ}{OP} + \frac{RP}{OP}.$$

If this last expression has its first term multiplied, both numerator and denominator, by OQ and the second term by QP, and the ratios rearranged, we get

$$\sin(x+y) = \frac{TQ}{OQ} \cdot \frac{OQ}{OP} + \frac{RP}{QP} \cdot \frac{QP}{OP}.$$

Using functions obtained from the figure, this may be written

$$\sin(x+y) = \sin x \cos y + \cos x \sin y.$$

From the same figure we have,

$$\cos(x+y) = \frac{OS}{OP} = \frac{OT-ST}{OP} = \frac{OT-RQ}{OP} = \frac{OT}{OP} - \frac{RQ}{OP}.$$

Multiplying, as we did above, the first fraction by OQ and the second fraction by QP, we get

$$\cos(x+y) = \frac{OT}{OQ} \cdot \frac{OQ}{OP} - \frac{RQ}{QP} \cdot \frac{QP}{OP}, \text{ or}$$

$$\cos(x+y) = \cos x \cos y - \sin x \sin y.$$

23. Sine and cosine of differences.

Using identity (VIII), the functions of negative angles of Art. 16, and assuming that the angle y may be negative, we have

$$\sin[x+(-y)] = \sin(x-y) = \sin x \cos(-y) + \cos x \sin(-y),$$

or

X. $\qquad \sin(x-y) = \sin x \cos y - \cos x \sin y.$

Similarly from identity (IX),

XI. $\qquad \cos(x-y) = \cos x \cos y + \sin x \sin y.$

24. Tangent of sum and difference.

Since $\tan(x+y) = \dfrac{\sin(x+y)}{\cos(x+y)}$, we have, using identities (VIII) and (IX),

$$\tan(x+y) = \frac{\sin x \cos y + \cos x \sin y}{\cos x \cos y - \sin x \sin y}.$$

Dividing numerator and denominator by cos x cos y, we have

$$\tan(x+y) = \frac{\dfrac{\sin x \cos y}{\cos x \cos y} + \dfrac{\cos x \sin y}{\cos x \cos y}}{\dfrac{\cos x \cos y}{\cos x \cos y} - \dfrac{\sin x \sin y}{\cos x \cos y}}.$$

Whence

XII. $$\tan(x+y) = \frac{\tan x + \tan y}{1 - \tan x \tan y}.$$

Similarly, we may show that

XIII. $$\tan(x-y) = \frac{\tan x - \tan y}{1 + \tan x \tan y}.$$

The identities of this assignment have important applications, and the student should study them carefully and memorize them. We proceed to illustrate some of their uses by examples.

EXAMPLE 1. Find the value of sin 75°.

SOLUTION: Since 75° = 45° + 30°, we have, using (VIII),

$$\sin 75° = \sin(45° + 30°) = \sin 45° \cos 30° + \cos 45° \sin 30°$$

$$= \frac{1}{\sqrt{2}} \cdot \frac{\sqrt{3}}{2} + \frac{1}{\sqrt{2}} \cdot \frac{1}{2} = \frac{\sqrt{3}+1}{2\sqrt{2}}.$$

EXAMPLE 2. Find the value of cos 15°.

SOLUTION: Since 15° = 45° − 30°, we have, using (XI),

$$\cos 15° = \cos(45° - 30°) = \cos 45° \cos 30° + \sin 45° \sin 30°$$

$$= \frac{1}{\sqrt{2}} \cdot \frac{\sqrt{3}}{2} + \frac{1}{\sqrt{2}} \cdot \frac{1}{2} = \frac{\sqrt{3}+1}{2\sqrt{2}}.$$

EXAMPLE 3. Show that cos (60° + x) + sin (30° + x) = cos x.

SOLUTION: Writing equivalent values for the left-hand member, we have

$$(\cos 60° \cos x - \sin 60° \sin x) + (\sin 30° \cos x + \cos 30° \sin x)$$

$$= \frac{1}{2} \cos x - \frac{\sqrt{3}}{2} \sin x + \frac{1}{2} \cos x + \frac{\sqrt{3}}{2} \sin x = \cos x.$$

Exercises

1. If $\sin x = \frac{2}{3}$ and $\cos y = \frac{4}{5}$ where x and y are acute angles, find $\sin(x+y)$, $\cos(x+y)$, $\tan(x+y)$.

2. Show that $\sin\left(\dfrac{\pi}{3} + x\right) - \cos\left(\dfrac{\pi}{3} + x\right) = \sin x$.

TANGENT OF SUM AND DIFFERENCE

3. If $\tan x = \frac{2}{5}$ and $\tan y = \frac{3}{2}$ where x and y are acute angles, find $\sin(x-y)$, $\cos(x-y)$, $\tan(x-y)$.
4. Find the values of sine, cosine and tangent of 15°.
5. Find the values of sine, cosine and tangent of 75°.
6. Using the identities of this assignment and known ratios, find

$\sin(45° + 60°)$ $\cos(180° + \theta)$ $\tan(180° + \theta)$

$\cos\left(\frac{\pi}{3} - \frac{\pi}{4}\right)$ $\sin(\pi - \theta)$ $\sin 120° = \sin(180° - 60°)$

$\sin(90° + \theta)$ $\cos(180° - \theta)$ $\cos 150° = \cos(180° - 30°)$

7. If x is a second quadrant angle with $\sin x = \frac{3}{5}$, and y is a first quadrant angle with $\cos y = \frac{2}{3}$, find $\sin(x+y)$ and $\cos(x-y)$.
8. (a) Express $\sin 19° \cos 43° + \cos 19° \sin 43°$ as the sine of a single angle.
   (b) Express $\cos 11° \cos 25° - \sin 11° \sin 25°$ as a function of a single angle.

9. Show that $$\tan(45° + x) = \frac{1 + \tan x}{1 - \tan x}.$$

10. Show that $$\frac{\tan(x+y) - \tan y}{1 + \tan(x+y)\tan y} = \tan x.$$

---ASSIGNMENT 10---

Double and Half Angles

25. Functions of double angles.

Suppose that in identity (VIII) of Art. 22, we have two equal angles, that is, suppose that $x = y$. Replacing y by x then gives

$$\sin(x + x) = \sin x \cos x + \cos x \sin x.$$

Hence

XIV. $\qquad \sin 2x = 2 \sin x \cos x.$

With the same assumptions, identity (IX) becomes

$$\cos(x + x) = \cos x \cos x - \sin x \sin x,$$

so that

XV. $\qquad \cos 2x = \cos^2 x - \sin^2 x.$

Two other forms of the relationship for $\cos 2x$ may be obtained by using identity (V), Art. 7, giving

$$\cos 2x = \cos^2 x - \sin^2 x = (1 - \sin^2 x) - \sin^2 x$$

or $\qquad \cos 2x = 1 - 2 \sin^2 x.$

Another form is,

$$\cos 2x = \cos^2 x - \sin^2 x = \cos^2 x - (1 - \cos^2 x)$$

or $\qquad \cos 2x = 2 \cos^2 x - 1.$

Finally, from (XII) we obtain

XVI. $\qquad \tan 2x = \dfrac{2 \tan x}{1 - \tan^2 x}.$

EXAMPLE 1. Find $\sin 60°$ from the functions of $30°$.

FUNCTIONS OF HALF ANGLES

SOLUTION: By (XIV), we have
$$\sin 2(30°) = \sin 60° = 2 \sin 30° \cos 30°$$
$$= 2 \cdot \frac{1}{2} \cdot \frac{\sqrt{3}}{2} = \frac{\sqrt{3}}{2}.$$

EXAMPLE 2. Show that $\sin 3x = 3 \sin x - 4 \sin^3 x$.

SOLUTION: Write $\sin 3x = \sin (2x + x)$ and apply (VIII).
$$\sin 3x = \sin 2x \cos x + \cos 2x \sin x.$$

Now apply identities (XIV) and (XV).
$$\sin 3x = 2 \sin x \cos x \cdot \cos x + (\cos^2 x - \sin^2 x) \sin x$$
$$= 2 \sin x \cos^2 x + \sin x \cos^2 x - \sin^3 x$$
$$= 3 \sin x \cos^2 x - \sin^3 x.$$

Now apply $\cos^2 x = 1 - \sin^2 x$, and obtain
$$\sin 3x = 3 \sin x(1 - \sin^2 x) - \sin^3 x$$
$$= 3 \sin x - 3 \sin^3 x - \sin^3 x$$
$$* \sin 3x = 3 \sin x - 4 \sin^3 x.$$

Exercises

1. Find the values of sine, cosine and tangent of 60°, using functions of 30°.
2. Find the values of sine, cosine and tangent of 120°, using functions of 60°.
3. Let $x = y$ in identities (X), (XI) and (XIII), and compare the results thus obtained with results previously established.
4. Prove that $\cos 3x = 4 \cos^3 x - 3 \cos x$.
5. Prove that $\dfrac{\sin 2x}{1 + \cos 2x} = \dfrac{1 - \cos 2x}{\sin 2x} = \tan x$.
6. Establish each of the following:
   (a) $\cos^4 x - \sin^4 x = \cos 2x$.   [*Hint:* Factor left member.]
   (b) $\sin 4x = 2 \sin 2x \cos 2x$.   [*Hint:* Let $4x = 2x + 2x$ or let $4x = 2(2x)$.]
   (c) $\dfrac{2 \tan x}{\tan 2x} = 1 - \tan^2 x$.

26. Functions of half angles.

By (V), Art. 7, and (XV), Art. 25, we have
$$\cos^2 x + \sin^2 x = 1$$
$$\cos^2 x - \sin^2 x = \cos 2x.$$

---

* This identity is often used as a basis of the proof of the fact that a general angle cannot be trisected by use of a straightedge and compasses alone. See Ball's *Mathematical Recreations*, p. 285.

DOUBLE AND HALF ANGLES

If we subtract the second identity from the first, we get
$$2 \sin^2 x = 1 - \cos 2x,$$
$$\sin^2 x = \frac{1 - \cos 2x}{2},$$
$$\sin x = \pm \sqrt{\frac{1 - \cos 2x}{2}}.$$

Now let $x = \frac{\theta}{2}$, and we get

XVII. $$\sin \frac{\theta}{2} = \pm \sqrt{\frac{1 - \cos \theta}{2}}.$$

By adding the two identities with which we began,
$$2 \cos^2 x = 1 + \cos 2x,$$
$$\cos x = \pm \sqrt{\frac{1 + \cos 2x}{2}},$$
and by letting $x = \frac{\theta}{2}$, we have

XVIII. $$\cos \frac{\theta}{2} = \pm \sqrt{\frac{1 + \cos \theta}{2}}.$$

The relation for the tangent function is

XIX. $$\tan \frac{\theta}{2} = \frac{\sin \frac{\theta}{2}}{\cos \frac{\theta}{2}} = \pm \sqrt{\frac{1 - \cos \theta}{1 + \cos \theta}}.$$

In each instance where a radical is concerned, we must choose the sign + or − according to the quadrant in which the angle $\frac{\theta}{2}$ lies, and in accord with Art. 13.

Exercises

1. Find the sine, cosine and tangent of 30° from the functions of 60°.
2. If $\cos \theta = \frac{4}{5}$, find $\sin \frac{\theta}{2}$, $\cos \frac{\theta}{2}$ and $\tan \frac{\theta}{2}$.

FUNCTIONS OF HALF ANGLES

3. Show that $\tan \dfrac{\theta}{2} = \dfrac{1 - \cos \theta}{\sin \theta}$.

Hint: Multiply both numerator and denominator of the quantity under the radical in (XIX) by $(1 - \cos \theta)$.

4. Show that $\tan \dfrac{\theta}{2} = \dfrac{\sin \theta}{1 + \cos \theta}$.

5. Show that $\cot \dfrac{\theta}{2} = \dfrac{1 + \cos \theta}{\sin \theta}$.

6.* Given $\tan 2\theta = -\tfrac{4}{3}$ and that 2θ terminates in the second quadrant, show that $\sin \theta$ and $\cos \theta$ have the values $\dfrac{2}{\sqrt{5}}$ and $\dfrac{1}{\sqrt{5}}$, respectively.

7. Given $\cos \dfrac{\theta}{2} = \dfrac{3}{5}$, find $\tan \theta$.

8. Show that $\cos 2x = \dfrac{1 - \tan^2 x}{1 + \tan^2 x}$.

* This and similar problems arise in analytic geometry. See Sisam's *Analytic Geometry*, p. 136.

ASSIGNMENT 11

Product Formulas

27. Product formulas.

We have already established the following formulas which we now rewrite for convenience.

$$\sin(x+y) = \sin x \cos y + \cos x \sin y \quad \text{(a)}$$
$$\sin(x-y) = \sin x \cos y - \cos x \sin y \quad \text{(b)}$$
$$\cos(x+y) = \cos x \cos y - \sin x \sin y \quad \text{(c)}$$
$$\cos(x-y) = \cos x \cos y + \sin x \sin y \quad \text{(d)}$$

If we let $x + y = A$ and $x - y = B$ and solve these two equations for x and y, we get

$$x = \tfrac{1}{2}(A+B), \quad y = \tfrac{1}{2}(A-B).$$

Substituting these values of x and y in the above four formulas, we get:

$$\sin A = \sin \tfrac{1}{2}(A+B) \cos \tfrac{1}{2}(A-B) + \cos \tfrac{1}{2}(A+B) \sin \tfrac{1}{2}(A-B) \quad \text{(e)}$$
$$\sin B = \sin \tfrac{1}{2}(A+B) \cos \tfrac{1}{2}(A-B) - \cos \tfrac{1}{2}(A+B) \sin \tfrac{1}{2}(A-B) \quad \text{(f)}$$
$$\cos A = \cos \tfrac{1}{2}(A+B) \cos \tfrac{1}{2}(A-B) - \sin \tfrac{1}{2}(A+B) \sin \tfrac{1}{2}(A-B) \quad \text{(g)}$$
$$\cos B = \cos \tfrac{1}{2}(A+B) \cos \tfrac{1}{2}(A-B) + \sin \tfrac{1}{2}(A+B) \sin \tfrac{1}{2}(A-B). \quad \text{(h)}$$

Add (e) to (f), and

XX. $\quad \sin A + \sin B = 2 \sin \tfrac{1}{2}(A+B) \cos \tfrac{1}{2}(A-B).$

Subtract (f) from (e), and

XXI. $\quad \sin A - \sin B = 2 \cos \tfrac{1}{2}(A+B) \sin \tfrac{1}{2}(A-B).$

Add (g) to (h), and

XXII. $\cos A + \cos B = 2 \cos \tfrac{1}{2}(A+B) \cos \tfrac{1}{2}(A-B).$

PRODUCT FORMULAS

Subtract (h) from (g), and

XXIII. $\cos A - \cos B = -2 \sin \tfrac{1}{2}(A+B) \sin \tfrac{1}{2}(A-B)$.

These last four identities are used in developing formulas for solving oblique triangles and will be discussed later.

EXAMPLE 1. Show that $\dfrac{\sin 41° - \sin 19°}{\cos 41° - \cos 19°} = -\sqrt{3}$.

SOLUTION: Using (XXI) for the numerator and (XXIII) for the denominator, where $A + B = 41° + 19° = 60°$, $A - B = 25°$,

$$\frac{\sin 41° - \sin 19°}{\cos 41° - \cos 19°} = \frac{2 \cos 30° \sin 12\tfrac{1}{2}°}{-2\sin 30° \sin 12\tfrac{1}{2}°} = -\cot 30° = -\sqrt{3}.$$

EXAMPLE 2. Show that $\dfrac{\sin 2x - \sin x}{\cos 2x + \cos x} = \tan \dfrac{x}{2} = \dfrac{\sin x}{1 + \cos x}$.

SOLUTION: Using (XXI) and (XXII) and the fact that
$$A + B = 2x + x = 3x, \quad A - B = x,$$

we have $\dfrac{\sin 2x - \sin x}{\cos 2x + \cos x} = \dfrac{2 \cos \tfrac{3x}{2} \sin \tfrac{x}{2}}{2 \cos \tfrac{3x}{2} \cos \tfrac{x}{2}} = \tan \dfrac{x}{2}.$

By Art. 26, Ex. 4, $\tan \dfrac{x}{2} = \dfrac{\sin x}{1 + \cos x}.$

Exercises

Express each of the following as a product.

1. $\sin 40° + \sin 30°$.   **3.** $\cos 5x - \cos 3x$.

2. $\cos 3x + \cos 5x$.   **4.** $\sin 6x - \sin 2x$.

Express each of the following as a sum or difference of functions.

5. $2 \sin 5x \cos 3x$.

Hint: $\tfrac{1}{2}(A+B) = 5x$, $\tfrac{1}{2}(A-B) = 3x$. Solve for A and B and use (XX).

6. $2 \sin \theta \cos \theta$. [Compare with (XIV), Art. 25.]

7. $2 \cos 3\theta \cos \theta$.   **8.** $\cos 5x \sin 3x$.

Prove each of the identities:

9. $\dfrac{\sin x + \sin 3x}{\cos x + \cos 3x} = \tan 2x$.

10.* $\dfrac{\sin A + \sin B}{\sin A - \sin B} = \dfrac{\tan \tfrac{1}{2}(A+B)}{\tan \tfrac{1}{2}(A-B)}$.

11. $\sin\left(x + \dfrac{\pi}{3}\right) + \sin\left(x - \dfrac{\pi}{3}\right) = \sin x$.   **12.** $\dfrac{\sin 4x - \sin 2x}{\cos 4x + \cos 2x} = \tan x$.

* This particular identity is used later in the solution of oblique triangles.

ASSIGNMENT 12

Logarithms

28. Logarithms.

The student has probably often expressed certain numbers as powers of 10. For example,

$$10000 = 10^4$$
$$1000 = 10^3$$
$$100 = 10^2.$$

It is possible to express any positive number as 10 with an exponent. One of the objects of this assignment is to show that we may write equations like the following:

$$500 = 10^{2.6990}$$
$$30 = 10^{1.4771}$$
$$5 = 10^{0.6990}$$

These equations are all of the form

$$N = 10^x \tag{1}$$

and it can be shown that for every positive number N there corresponds a definite value x. The exponent x is called the **common logarithm** of N or logarithm of N to the base 10, and the equation $N = 10^x$ may be written

$$\log N = x. \tag{2}$$

Equations (1) and (2) give the same information. The first is called the exponential form, the second the logarithmic form.

Bases other than 10 may be used in connection with logarithmic theory, but the base 10 is the most convenient for most computations.

LAWS OF LOGARITHMS

Exercises

1. Write the following equations in the logarithmic form; thus if $17 = 10^{1.2304}$, then $\log 17 = 1.2304$.
   - (a) $100000 = 10^5$
   - (b) $1700 = 10^{3.2304}$
   - (c) $23 = 10^{1.3617}$
   - (d) $0.1 = 10^{-1}$
   - (e) $0.743 = 10^{-0.1290}$

2. Write the following equations in the exponential form; thus if
   $$\log 25 = 1.3979, \text{ then } 25 = 10^{1.3979}.$$
   - (a) $\log 3 = 0.4771$
   - (b) $\log 27 = 1.4314$
   - (c) $\log 1830 = 3.2625$
   - (d) $\log 0.01 = -2$
   - (e) $\log 0.143 = 9.1553 - 10$

29. Laws of logarithms.

There are three fundamental laws used in operations with logarithms.

Law 1. *The logarithm of a product equals the sum of the logarithms of its factors.* In symbols this law is written

$$\log AB = \log A + \log B.$$

To prove this law let

$$\log A = x \text{ and } \log B = y$$

then $\qquad A = 10^x \text{ and } B = 10^y$

and $\qquad A \cdot B = 10^x \cdot 10^y = 10^{x+y}.$

Hence $\qquad \log AB = x + y = \log A + \log B.$

In a similar way we may show that

$$\log ABC = \log A + \log B + \log C$$

and so on for any number of factors.

Law 2. *The logarithm of a quotient is equal to the logarithm of the dividend minus the logarithm of the divisor.* In symbols,

$$\log \frac{A}{B} = \log A - \log B.$$

To prove, let

$$\log A = x, \log B = y$$

then $\quad A = 10^x, \quad B = 10^y$

and $\quad \dfrac{A}{B} = \dfrac{10^x}{10^y} = 10^{x-y}.$

Hence, $\quad \log \dfrac{A}{B} = x - y = \log A - \log B.$

Law 3. $\log A^r = r \log A.$

To prove, let $\quad \log A = x,$

then $\quad A = 10^x$

and $\quad A^r = (10^x)^r = 10^{xr}.$

Hence $\quad \log A^r = xr = r \log A.$

COROLLARY. *The logarithm of the real positive nth root of a number is the logarithm of the number divided by n.* In symbols,

$$\log \sqrt[n]{A} = \log A^{\frac{1}{n}} = \frac{\log A}{n}$$

Exercises

1. Given $\log 2 = 0.3010$, $\log 3 = 0.4771$, $\log 7 = 0.8451$, $\log 10 = 1$, find

(a) $\log 6$
(b) $\log 42$
(c) $\log 5$
(d) $\log 49$
(e) $\log \frac{7}{2}$
(f) $\log \sqrt{3}$
(g) $\log \sqrt[3]{7}$
(h) $\log (14)^5$
(i) $\log (420)^2$
(j) $\log 10.5$

Express the logarithms of the following in terms of the logarithms of integers:

2. $\dfrac{\sqrt{2}}{\sqrt[3]{3}}.$

SOLUTION: $\quad \log \dfrac{\sqrt{2}}{\sqrt[3]{3}} = \log \sqrt{2} - \log \sqrt[3]{3}$

$\qquad\qquad\qquad\quad = \tfrac{1}{2} \log 2 - \tfrac{1}{3} \log 3.$

3. $5^3 \cdot \sqrt{7}$

4. $3^{\frac{2}{3}} \cdot 7^{\frac{1}{5}} \cdot 9^2$

5. $\dfrac{\sqrt[3]{9}}{\sqrt{6} \cdot \sqrt[4]{10}}$

6. $8^5 \cdot 7^3 \cdot 4^{\frac{1}{3}}$

Express the logarithms of the following in terms of the logarithms of prime numbers.

7. 210

8. $\sqrt{75}$

9. $\sqrt[3]{100}$

10. $\frac{36}{81}$

11. $\dfrac{(25)^{\frac{1}{4}}}{(36)^{\frac{1}{3}}}$

12. $\sqrt{9.24}$

ASSIGNMENT 13

Logarithms, Use of Tables

30. Characteristic and mantissa.

A logarithm consists of two parts, an integral part and a decimal part. For example, as we shall show later, log 597 = 2.7627. The integral part of a logarithm is called the **characteristic** and the decimal part the **mantissa**. The characteristic may be negative. It is usually convenient to keep the mantissa positive even if the logarithm is negative. For example,

$$\log \tfrac{2}{3} = \log 2 - \log 3 = 0.3010 - 0.4771 = -0.1761,$$

but this may be written $-1 + 0.8239$ or $\bar{1}.8239$ which has a positive mantissa and negative characteristic. However, we shall find it more convenient to write this logarithm $9.8239 - 10$.

We may write the following series of equalities

$$\begin{aligned} 10^3 &= 1000, \text{ or } \log 1000 = 3, \\ 10^2 &= 100, \text{ or } \log 100 = 2, \\ 10^1 &= 10, \text{ or } \log 10 = 1, \\ 10^0 &= 1, \text{ or } \log 1 = 0. \end{aligned}$$

. . . .

Note that all numbers between 1000 and 100 are numbers with three digits in the integral part, for example, 937, 536.7, 247.3. The logarithms of these numbers will have 2 for a characteristic. Looking over the above series of equalities we may formulate —

RULE I. *The characteristic of the logarithm of a number greater than unity is one less than the number of digits in the integral part of the number.*

For numbers less than unity we may write

$10^0 = 1$, or $\log 1 = 0$
$10^{-1} = 0.1$ or $\log 0.1 = -1 = 9.0000 - 10$
$10^{-2} = 0.01$ or $\log 0.01 = -2 = 8.0000 - 10$
$10^{-3} = 0.001$ or $\log 0.001 = -3 = 7.0000 - 10$
. . . .

From this series of equalities we formulate —

RULE II. *To find the characteristic of the common logarithm of a number between 0 and 1, subtract from 9 the number of ciphers between the decimal point and the first significant figure. From the number so obtained subtract 10.*

For example, to find the characteristic of the logarithm of 0.0384 we find one cipher between the decimal point and 3, the first significant figure. The characteristic is then $8 - 10$, or -2. As we shall show later, the mantissa is .5834 and we may write $\log 0.0384 = \bar{2}.5834$ or preferably $8.5834 - 10$.

A negative number has no real logarithm.

Consider the following sequence of logarithmic statements:

$\log 288 = 2.4594$,
$\log 2880 = \log(288 \cdot 10) = \log 288 + \log 10 = 2.4594 + 1 = 3.4594$
$\log 28.8 = \log \frac{288}{10} = \log 288 - \log 10 = 2.4594 - 1 = 1.4594$.
$\log 0.288 = \log \frac{288}{1000} = \log 288 - \log 1000 = 2.4594 - 3 = 9.4594 - 10$

We note that the mantissa is unchanged if the number contains the same sequence of significant digits, and that only the characteristic changes when the decimal point is moved.

Hence, tables of common logarithms contain only the mantissas, the characteristics being found in accord with Rules I and II.

31. To find the logarithm of a number from the table.

On pages 106–107 there is a four-place table of logarithms. Such a table gives the mantissas to four decimal places. In

INTERPOLATION

the left-hand column of this table under n are found the first two digits of the number whose logarithm is to be found. The third digit is found in the black type in the top row of the table indicating the column in which the mantissa is to be found. For example, to find the logarithm of 738, we locate 73 in the column under n on page 107. Moving across the page to the column headed 8, we find 8681 for the mantissa. Applying the rules for characteristics, we have log 738 = 2.8681.

Exercises

1. Find log 341.
2. Find log 57.6.
3. Find log 9.99.
4. Find log 1.34.
5. Find log 0.745.
6. Find log 0.0111.
7. Find log 71600.
8. Find log 1010000.
9. Find log 0.00062.
10. Find log 2490.

32. Interpolation.

From the table of logarithms, we can obtain directly the mantissa of the logarithm of any three-figure number. If the number has four or more significant figures, its logarithm is not recorded directly in the table but may be obtained approximately by a process of **interpolation**. In this process, it is assumed that to a small change in the number, there corresponds a change in the logarithm which is proportional to the change in the number. This assumption is called the **principle of proportional parts**.

EXAMPLE. Find the logarithm of 28.37.
From the table we find,
$$\log 28.30 = 1.4518$$
$$\log 28.40 = 1.4533.$$

The number 28.27 lies .7 of the way from 28.30 to 28.40. Hence by the principle of proportional parts the mantissa of the logarithm of 28.37 will lie .7 of the way from 4518 to 4533. The difference is 15, hence we add .7 × 15 = 10.5 to 4815, to get log 28.37 = 1.4528.*

---

* Since .7 × 15 = 10.5 is halfway between 10 and 11 we can say that log 28.37 is either 1.4528 or 1.4529 when written to four decimal places. It is an old computer's rule, where such a choice is given to take the result ending in an even number. If this rule is followed in a long piece of computation, the errors tend to compensate one another.

54 LOGARITHMS, USE OF TABLES

Exercises
Find from the table the logarithms of the following numbers:

1. 2713	**3.** 999.9	**5.** 3.445	**7.** 481300	**9.** 0.09832
2. 793.4	**4.** 617.7	**6.** 0.5231	**8.** 0.007326	**10.** 27137

33. To find from the table the number which corresponds to a given logarithm.

EXAMPLE 1. Find the number whose logarithm is 1.8274.

Looking in the table we find the mantissa 8274 in the row which has 67 in the n column and in the column which has 2 at the top. The sequence of digits corresponding to the mantissa 8274 is then 672, and the rule for characteristics places the decimal point between the 7 and the 2, and

$$\log 67.2 = 1.8274.$$

EXAMPLE 2. Find the number whose logarithm is 2.4823.

The mantissa 4823 is not found in the body of the table, but lies between 4814 and 4829. Since we have

$$4829 - 4814 = 15$$
and
$$4823 - 4814 = 9$$

the number 4823 lies $\frac{9}{15}$ of the way from 4814 to 4829. From the table we find

$$\log 303 = 2.4814$$
$$\log N \; = 2.4823$$
$$\log 304 = 2.4829$$

where N is the number we are looking for. Interpolation by proportional parts gives N, $\frac{9}{15}$ of the way from 303 to 304 or 303.6. Hence $\log 303.6 = 2.4823$.

Exercises
Find the numbers whose logarithms are the following:

1. 0.3876	**3.** 1.9470	**5.** 3.7871
2. 2.5547	**4.** 9.9140 − 10	**6.** 4.9296
7. 8.6789 − 10		**9.** 3.0271
8. 7.4335 − 10		**10.** 2.0093

ASSIGNMENT 14

Logarithms (Applications)

34. Use of logarithms in computation.

The application of logarithms to computation depends on the laws of logarithms given on pages 49–50, as shown in the following examples.

EXAMPLE 1. Find the value of $N = \dfrac{2.673}{3.252}$ to four significant figures.

SOLUTION:
$$\log 2.673 = 0.4270$$
$$\log 3.252 = 0.5122,$$
$$\log N = \log 2.673 - \log 3.252 = -0.0852 = 9.9148 - 10$$
$$N = 0.8218.$$

EXAMPLE 2. Find $N = \dfrac{\sqrt{84.32}}{\sqrt[3]{84.32}}$ to four significant figures.

SOLUTION:
$$\log \sqrt{84.32} = \tfrac{1}{2} \log 84.32 = \tfrac{1}{2}(1.9259)$$
$$= 0.9630$$
$$\log \sqrt[3]{84.32} = \tfrac{1}{3}(1.9259) = 0.6420$$
$$\log N = 0.9630 - 0.6420 = 0.3210$$
$$N = 2.094.$$

EXAMPLE 3. Find the fifth root of 5.555, to four significant figures.

SOLUTION:
$$\log \sqrt[5]{5.555} = \tfrac{1}{5} \log 5.555$$
$$= \frac{0.7447}{5} = 0.1489$$
$$\sqrt[5]{5.555} = 1.409$$

EXAMPLE 4. Find the fifth root of 0.0555 to four significant figures.

56 LOGARITHMS (APPLICATIONS)

SOLUTION: $\log \sqrt[5]{0.05555} = \frac{1}{5} \log 0.05555$

$$= \frac{8.7447 - 10}{5} = \frac{48.7447 - 50}{5} = 9.7489 - 10$$

$$\sqrt[5]{0.05555} = 0.5609.$$

EXAMPLE 5. One thousand dollars is allowed to accumulate at 5 per cent interest compounded annually. What is the amount at the end of ten years?

SOLUTION: At the end of 10 years the amount is $1000 \cdot (1.05)^{10}$.

$$\log 1000 \cdot (1.05)^{10} = \log 1000 + 10 \cdot \log 1.05$$
$$= 3 + 10 \cdot (.0212)$$
$$= 3.212$$
$$1000 \cdot (1.05)^{10} = 1629. \text{ Ans. } \$1629.$$

Exercises

Compute to four significant figures by the use of logarithms.

1. $(38.24) \cdot (973.2)$
2. $(7.136) \cdot (81.79)(678.9)$
3. $(0.8417) \cdot (0.03174)$
4. $\dfrac{7.634}{6.271}$
5. $\dfrac{6.271}{7.634}$

6. $\dfrac{0.8317}{0.7132}$
7. $\dfrac{8.888}{0.007654}$
8. $\dfrac{1}{9876}$
9. $(6.271)^5$
10. $(3.999)^3 \sqrt{11.49}$

11. $\sqrt[7]{1.234}$
12. $\sqrt[7]{0.01234}$
13. $(0.7555)^5$
14. $\left(\dfrac{3}{4}\right)^{10}$
15. $\left(\dfrac{1}{3}\right)^{\frac{1}{3}}$

16. $(0.25)^{0.25}$

SOLUTION: $\log 0.25 = 9.3979 - 10$
$\log (0.25)^{0.25} = 0.25(9.3979 - 10)$
$= 2.3495 - 2.5$
$= -0.1505$
$= 9.8495 - 10$
$(0.25)^{0.25} = 0.7072$

17. $(0.5)^{0.7}$
18. $\dfrac{\sqrt{3}}{\sqrt[3]{4}}$
19. $(0.51)^{-0.09}$
20. $\dfrac{(0.6741)^{\frac{1}{3}}}{(12.36)^{\frac{1}{2}}}$

ASSIGNMENT 15

Solution of Triangles using Logarithms

35. Right triangles.

In Art. 10 we have solved right triangles using ordinary arithmetic and tables of values of the trigonometric functions. In this section we solve right triangles using tables of the logarithms of trigonometric functions (Table III, page 113), and then go on to the solution of oblique triangles. The solution of right triangles is best shown by illustrative examples.

EXAMPLE 1. In the triangle in Fig. 33 the hypotenuse c is 23.34 feet and the angle A is 33° 7'. Find a, b, and B.

SOLUTION: $B = 90° - 33° 7' = 56° 53'$.

From the definitions of the trigonometric functions we have

$$\frac{a}{c} = \sin A, \quad \text{or} \quad a = c \sin A.$$

$$\frac{b}{c} = \cos A, \quad \text{or} \quad b = c \cos A.$$

We may then write

$\log a = \log c + \log \sin A$
$\log b = \log c + \log \cos A$

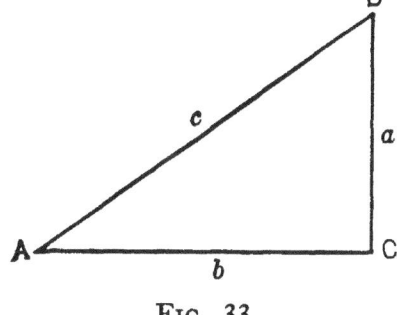

FIG. 33

and the computation is arranged as follows:

$\log c = \log 23.34 \quad\quad = 1.3681$
$\log \sin A = \log \sin 33° 7' = 9.7374 - 10$ (Found on page
adding $\quad\quad\quad\quad\quad \log a = \overline{1.1055} \quad\quad$ 116 of Table III)
$\quad\quad\quad\quad\quad\quad\quad\quad a = 12.75$

SOLUTION OF TRIANGLES

$$\log c = 1.3681$$
$$\log \cos A = \log \cos 33° 7' = 9.9230 - 10$$
adding
$$\log b = \overline{1.2911}$$
$$b = 19.55$$

In interpolation in the cosine column of the table it should be noted that the cosine decreases as the angle increases.

NOTE. In this example we have written down the -10 which comes with the logarithms of sines and cosines. This is unnecessary, and hereafter we shall omit it.

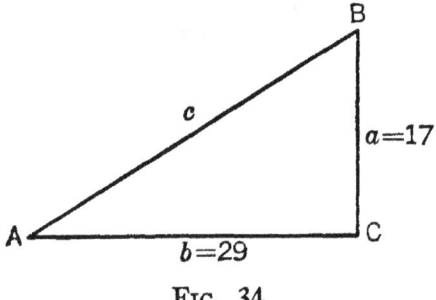

FIG. 34

EXAMPLE 2. In the triangle in Fig. 34, $a = 17$, $b = 29$, find A, B, c.

Using the formula $\tan A = \dfrac{a}{b}$ we have

$$\log a = \log 17 = 1.2304 = (11.2304 - 10)$$
$$\log b = \log 29 = 1.4624$$
subtracting
$$\log \tan A = \overline{9.7680}$$
$$A = 30° 22'$$
$$B = 90° - A = 59° 38'$$

To find c we can use either $c = \dfrac{a}{\sin A}$, or $c = \sqrt{a^2 + b^2}$

$$\log a = \log 17 = 1.2304$$
$$\log \sin A = \log \sin 30° 22' = 9.7037$$
subtracting
$$\log c = \overline{1.5267}$$
$$c = 33.63$$

Using
$$c = \sqrt{a^2 + b^2} = \sqrt{17^2 + 29^2} = \sqrt{1130}$$
$$\log 1130 = 3.0531$$
$$\log \sqrt{1130} = 1.5266 = \log c$$
$$c = 33.62$$

NOTE. In computations with four-place tables the fourth significant figure is doubtful. As in this example solutions by two different methods may differ in the fourth place. If four-figure accuracy is wanted, five-place tables should be used.

RIGHT TRIANGLES

Exercises

Solve the following right triangles:

1. $c = 120, A = 31°$
2. $c = 93.4, A = 13° 35'$
3. $a = 637, A = 4° 35'$
4. $a = 731, A = 1$ radian
5. $a = 2.189, B = 45° 25'$
6. $a = 18.26, B = 36° 30'$
7. $b = 16.93, B = 51° 2'$
8. $b = 14.95, B = 1.5$ radians
9. $b = 7333, A = \frac{1}{2}$ radian
10. $b = 50.94, A = 46° 12'$
11. $a = 0.7183, c = 1.634$
12. $a = 8.714, c = 9.914$
13. $a = 36, b = 83$
14. $a = 213.4, b = 928.3$. (Find angles in radians.)
15. $b = 0.0039, c = 0.0058$. (Find angles in radians.)
16. $b = 10, c = 341$
17. At a horizontal distance of 253 feet from the foot of a tower the angle of elevation of the top was found to be 37° 26'. Find the height of the tower.
18. How high is a tree which casts a shadow 83 feet long when the angle of elevation of the sun is 49° 32'? (See page 21.)
19. From the top of a vertical cliff 347 feet high, the angle of depression of a boat was found to be 25° 14'. Find the distance of the boat from the foot of the cliff.
20. A tower on a level plain is 179 feet high. A surveyor measures the angle of elevation of the top and finds it to be 4° 3'. How far is he from the tower?

ASSIGNMENT 16

Solution of Triangles

36. Oblique triangles.

If three of the six parts of a triangle are given, the other three parts can be found, provided the three parts are not all angles and assuming the data to be such that a triangle is possible.

This gives rise to four cases:

Case I. Given one side and two angles.
Case II. Given two sides and an angle opposite one of them.
Case III. Given two sides and the included angle.
Case IV. Given the three sides.

37. Law of sines.

Consider any triangle ABC (Figs. 35, 36).

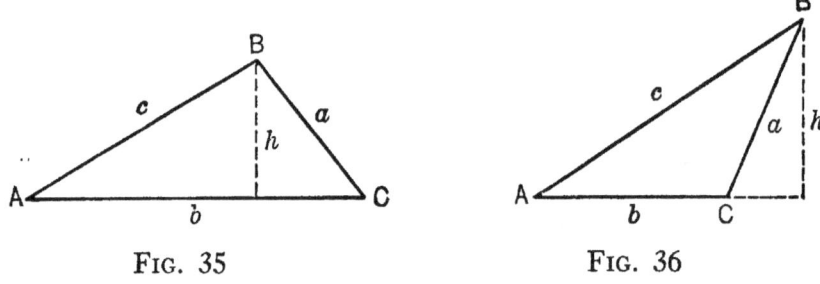

FIG. 35   FIG. 36

The perpendicular from a vertex to the opposite side is denoted by h. From either triangle we have

$$\frac{h}{c} = \sin A, \qquad \frac{h}{a} = \sin C$$

GIVEN A SIDE AND TWO ANGLES   61

if we recall that sin $(180° - c)$ = sin C. Solving each of these equations for h and equating the values, we have

$$c \sin A = a \sin C$$

or
$$\frac{a}{\sin A} = \frac{c}{\sin C}$$

In the same way, by dropping perpendiculars from A to the opposite side, or the opposite side produced, we get

$$\frac{b}{\sin B} = \frac{c}{\sin C}$$

and we may write

$$\frac{a}{\sin A} = \frac{b}{\sin B} = \frac{c}{\sin C}$$

which is the **law of sines**. Written in words the law of sines is:

In any triangle the sides are proportional to the sines of the opposite angles.

38. Case I. Given a side and two angles.

EXAMPLE 1. Given $b = 64.25$, $A = 24° 23'$, $B = 61° 48'$. Find a, c and C.

SOLUTION: Since the sum of the angles of any triangle is 180°, we find $C = 180° - (A + B) = 93° 49'$.

To find a, we use $\dfrac{a}{\sin A} = \dfrac{b}{\sin B}$ or $a = \dfrac{b \sin A}{\sin B}$

or
$$\log a = \log b + \log \sin A - \log \sin B$$

```
       log b = log 64.25            = 1.8078
   log sin A = log sin 24° 23'      = 9.6157
      adding                          1.4235
   log sin B = log sin 61° 48'      = 9.9452
   subtracting              log a  = 1.4783
                            a      = 30.08.
```

In a similar way, using $\dfrac{c}{\sin C} = \dfrac{b}{\sin B}$, or $c = \dfrac{b \sin C}{\sin B}$, we find c.

```
                        log b = log 64.25            = 1.8078
log sin 93° 49' = log sin (180° - 86° 11') = log sin 86° 11' = 9.9990
                       adding                          1.8068
                   log sin B = log sin 61° 48'      = 9.9452
                   subtracting              log c  = 1.8616
                                            c      = 72.72.
```

SOLUTION OF TRIANGLES

EXAMPLE 2. Given $a = 259.4$, $A = 47° 10'$, $C = 59° 17'$, find b, c, B.

SOLUTION: $B = 180° - (A + B) = 73° 33'$.

Using $\dfrac{b}{\sin B} = \dfrac{a}{\sin A}$ and $\dfrac{c}{\sin C} = \dfrac{a}{\sin A}$, we have

log a = 2.4140	log a = 2.4140
log sin B = 9.9818	log sin C = 9.9344
2.3958	2.3484
log sin A = 9.8653	log sin A = 9.8653
log b = 2.5305	log c = 2.4831
b = 339.2	c = 304.1

Exercises and Problems

Solve the following triangles, given

1. $A = 47° 13'$, $B = 65° 24'$, $a = 43.18$.
2. $B = 65° 15'$, $C = 81° 25'$, $a = 43.82$.
3. $A = 65° 50'$, $B = 38° 37'$, $b = 835.6$.
4. $B = 115° 34'$, $C = 15° 57'$, $b = 5.44$.
5. $A = 68° 41'$, $B = 1° 2'$, $c = 9.433$.
6. $B = 17° 57'$, $C = 78° 15'$, $c = 1622$.
7. $a = 3541$, $A = 61° 27'$, $C = 33° 22'$.
8. $b = 4017$, $A = 61° 27'$, $C = 33° 22'$.

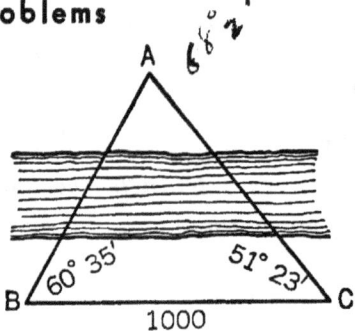

FIG. 37

9. To determine the distance of the inaccessible point A from B, a point C was marked 1000 yards from B. The angles ABC and BCA were measured and found to be $60° 35'$ and $51° 23'$. Find the distance AB.

10. The angle of elevation of the top of a tower is observed to be $13° 6'$. At a point 148.6 feet nearer to the tower the angle of elevation is $19° 17'$. Find the height of the tower.

―――――ASSIGNMENT 17―――――

Solution of Triangles (Case II)

39. Case II. Two sides and the angle opposite one of them.

This case is often called the "ambiguous" case, for there may be two solutions, one solution or no solution. If we consider the given parts to be a, c, and A, there will be one solution if the conditions shown in Figs. 38, 39, and 40 are satisfied.

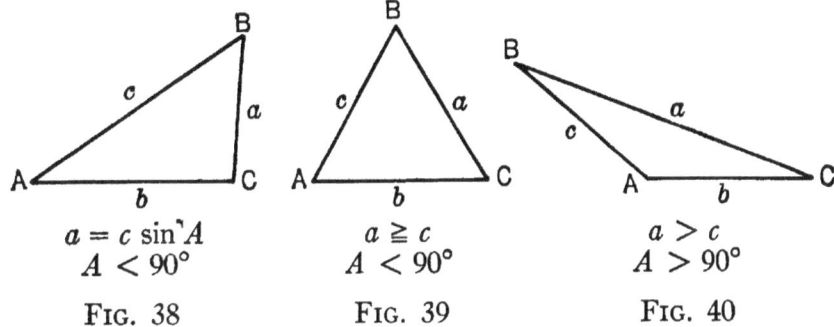

$a = c \sin A$ $\qquad$ $a \geqq c$ $\qquad\qquad$ $a > c$
$A < 90°$ $\qquad\quad$ $A < 90°$ $\qquad\quad$ $A > 90°$

FIG. 38 $\qquad\qquad$ FIG. 39 $\qquad\qquad$ FIG. 40

If the side a is too short to reach the line b, there is no solution, as shown in Figs. 41, 42.

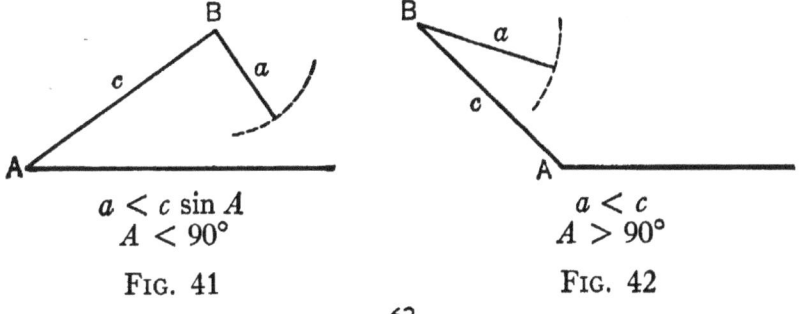

$a < c \sin A$ $\qquad\qquad$ $a < c$
$A < 90°$ $\qquad\qquad\quad$ $A > 90°$

FIG. 41 $\qquad\qquad\qquad$ FIG. 42

64 SOLUTION OF TRIANGLES (CASE II)

Lastly, there may be two solutions (Fig. 43).

EXAMPLE 1. Given $a = 531$, $c = 835$, $A = 41° 8'$, find b, B, C.

SOLUTION: Using $\sin C = \dfrac{c \sin A}{a}$

adding
$$\begin{aligned}\log c &= 2.9217 \\ \log \sin A &= 9.8181 \\ \hline &2.7398 \\ \log a &= 2.7251 \\ \hline \log \sin C &= .0147\end{aligned}$$

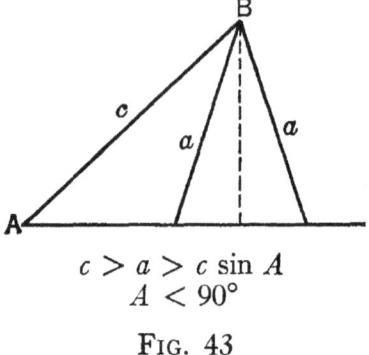

$c > a > c \sin A$
$A < 90°$

FIG. 43

which means that $\sin C$ is greater than 1, which is impossible, hence there is no solution.

EXAMPLE 2. Given $a = 549.2$, $c = 835$, $A = 41° 8'$, find b, B, C.

SOLUTION:
$$\begin{aligned}\log c &= 2.9217 \\ \log \sin A &= 9.8181 \\ \hline &2.7398 \\ \log a &= 2.7398 \\ \hline \log \sin C &= .0000 \\ C &= 90°.\end{aligned}$$

This is the case in Fig. 38, hence there is one solution.
$$B = 180° - (A + C) = 48° 52'.$$

Using $b = \dfrac{a \sin B}{\sin A}$
$$\begin{aligned}\log a &= 2.7398 \\ \log \sin B &= 9.8769 \\ \hline &2.6167 \\ \log \sin A &= 9.8181 \\ \hline \log b &= 2.7986 \\ b &= 628.9.\end{aligned}$$

EXAMPLE 3. Given $a = 840$, $c = 835$, $A = 41° 8'$, find b, B, C.

SOLUTION:
$$\begin{aligned}\log c &= 2.9217 \\ \log \sin A &= 9.8181 \\ \hline &2.7398 \\ \log a &= 2.9243 \\ \hline \log \sin C &= 9.8155 \\ C &= 40° 50' \\ B = 180° - (A + C) &= 98° 2'\end{aligned}$$

This is the case of Fig. 39.

Using $b = \dfrac{a \sin B}{\sin A}$,

TWO SIDES AND ANGLE OPPOSITE ONE

$$\log a = 2.9243$$
$$\log \sin B = 9.9958$$
$$\overline{2.9201}$$
$$\log \sin A = 9.8181$$
$$\log b = \overline{3.1020}$$
$$b = 1265.$$

EXAMPLE 4. Given $a = 625$, $c = 835$, $A = 41°\,8'$, find b, B, C.

A rough preliminary sketch shows two solutions as in Fig. 44. Or we compare a with $c \sin A$ as shown below.

$$\log c = 2.9217$$
$$\log \sin A = \underline{9.8181}$$
$$\log c \sin A = 2.7398 \quad \text{Note here that}$$
$$\log a = 2.7959 \quad \log a > \log c \sin A$$
$$\log \sin C = \overline{9.9439} \quad \text{or } a > c \sin A.$$
$$C = 61°\,30'$$

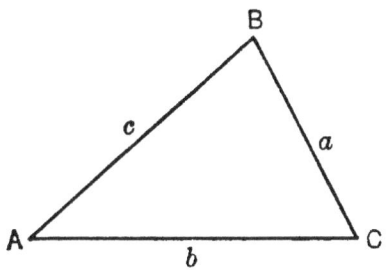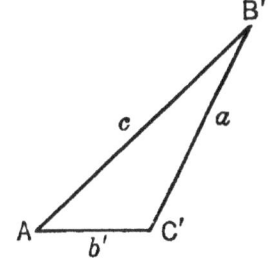

FIG. 44

This gives the angle at C (Fig. 44). $B = 77°\,22'$.

To find side b

$$\log a = 2.7959$$
$$\log \sin B = 9.9894$$
$$\overline{2.7853}$$
$$\log \sin A = 9.8181$$
$$\log b = \overline{2.9672}$$
$$b = 927.2.$$

The angle $C' = 180 - C = 118°\,30'$.

From the second triangle $B' = 180° - (A + C') = 20°\,22'$.

$$\log a = 2.7959$$
$$\log \sin B' = 9.5416$$
$$\overline{2.3375}$$
$$\log \sin A = 9.8181$$
$$\log b' = \overline{2.5194}$$
$$b' = 330.7$$

SOLUTION OF TRIANGLES (CASE II)

Exercises and Problems

Solve the following triangles, given:
1. $a = 443, c = 439, A = 40° 12'$.
2. $a = 43.2, b = 53.5, A = 47° 13'$.
3. $b = 724.7, c = 787.5, B = 65° 15'$.
4. $b = 724.7, c = 787.5, C = 65° 15'$.
5. $a = 1149, b = 1246, A = 67° 16'$.
6. $a = 251.2, b = 222.2, B = 59° 27'$.
7. $a = 3541, b = 4017, A = 61° 27'$.
8. $b = 769.8, c = 3216, C = 73° 11'$.
9. $a = 1000, c = 2000, C = 30°$.
10. $a = 0.2111, b = 0.5731, A = 46° 46'$.
11. Given $a = 712, b = 296, B = 16° 16'$, find the difference between the lengths of the sides c of the two triangles.

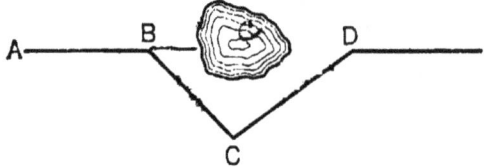

Fig. 45

12. A surveyor is prevented from continuing the line AB (Fig. 45) by an obstruction. He then runs a line BC 300 feet long, making the angle $ABC = 137°$. He then runs a line $CD = 375.7$ feet long. What was the angle BCD and the distance BD?

────────ASSIGNMENT 18────────

Solution of Triangles (Case III)

40. Law of cosines.

For each of the triangles in Figs. 46, 47, 48, we have

$$a^2 = h^2 + \overline{PC}^2, \tag{1}$$
$$c^2 = h^2 + \overline{AP}^2. \tag{2}$$

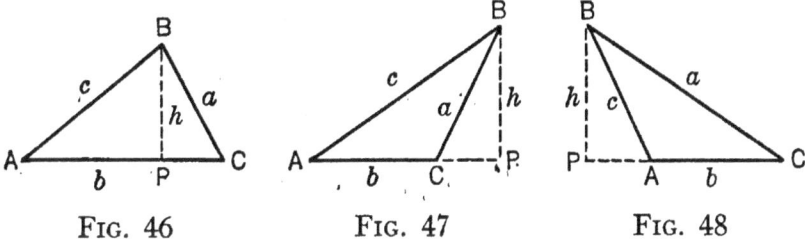

FIG. 46   FIG. 47   FIG. 48

Let the second equation be solved for h^2 and the result substituted in the first equation. We get

$$a^2 = c^2 - \overline{AP}^2 + \overline{PC}^2. \tag{3}$$

Note that PC is negative in Fig. 47 and AP is negative in Fig. 48 so we may write

$$PC = b - AP$$

for all three figures, and equation (3) becomes

$$a^2 = c^2 - \overline{AP}^2 + (b - AP)^2$$
$$= c^2 - \overline{AP}^2 + b^2 - 2b\overline{AP} + \overline{AP}^2$$
$$= c^2 + b^2 - 2b\overline{AP},$$

but
$$\overline{AP} = c \cos A,$$
hence
$$a^2 = c^2 + b^2 - 2bc \cos A. \tag{4}$$
In a similar way we can show that
$$b^2 = a^2 + c^2 - 2ac \cos B, \tag{5}$$
$$c^2 = a^2 + b^2 - 2ab \cos C. \tag{6}$$

These equations give the **law of cosines** which put into words is:

The square of any side of a triangle is equal to the sum of the squares of the other two sides minus twice their product multiplied by the cosine of the included angle.

41. Law of tangents. From the law of sines, we have
$$\frac{\sin A}{\sin B} = \frac{a}{b}, \quad \frac{\sin C}{\sin B} = \frac{c}{b}.$$

Adding, we have
$$\frac{\sin A}{\sin B} + \frac{\sin C}{\sin B} = \frac{a}{b} + \frac{c}{b},$$
or
$$\frac{\sin A + \sin C}{\sin B} = \frac{a+c}{b}. \tag{7}$$

Subtraction gives
$$\frac{\sin A - \sin C}{\sin B} = \frac{a-c}{b}. \tag{8}$$

Dividing (8) by (7) we find
$$\frac{\sin A - \sin C}{\sin A + \sin C} = \frac{a-c}{a+c}. \tag{9}$$

From Art. 27 we know
$$\sin A - \sin C = 2 \cos \frac{A+C}{2} \sin \frac{A-C}{2},$$
$$\sin A + \sin C = 2 \sin \frac{A+C}{2} \cos \frac{A-C}{2},$$
or
$$\frac{\sin A - \sin C}{\sin A + \sin C} = \frac{\tan \frac{1}{2}(A-C)}{\tan \frac{1}{2}(A+C)} = \frac{a-c}{a+c}. \tag{10}$$

In a similar way we find
$$\frac{\tan \frac{1}{2}(B-C)}{\tan \frac{1}{2}(B+C)} = \frac{b-c}{b+c}, \quad \frac{\tan \frac{1}{2}(A-B)}{\tan \frac{1}{2}(A+B)} = \frac{a-b}{a+b}. \tag{11}$$

TWO SIDES AND INCLUDED ANGLE

These equations express the **law of tangents.** In words the law of tangents is:

In any triangle the difference of two sides is to their sum as the tangent of half the difference of the opposite angles is to the tangent of half their sum.

42. Case III. Given two sides and the included angle.

Either the law of cosines or the law of tangents can be used in solving this case. The law of cosines is not so well adapted to logarithmic calculations as is the law of tangents. However, if the sides of the triangle are small numbers, or are given only to one or two significant figures, or if only one side is to be found, the law of cosines will be found useful.

EXAMPLE 1. Given $A = 47°$, $b = 11$, $c = 17$. Find a.

SOLUTION: From the law of cosines we have
$$a^2 = b^2 + c^2 - 2bc \cos A$$
$$= 121 + 289 - 2 \cdot 11 \cdot 17 \cdot .6820$$
$$= 154.9.$$
$$a = 12.4.$$

EXAMPLE 2. Given $A = 47° 18'$, $b = 11.34$, $c = 17.28$, find a, B, C.

SOLUTION: From the law of tangents, we have
$$\tan \frac{1}{2}(C - B) = \frac{c - b}{c + b} \tan \frac{1}{2}(C + B).$$

$c - b = 5.94$
$c + b = 28.62$
$C + B = 180° - A = 132° 42'$, $\frac{1}{2}(C + B) = 66° 21'$

$\log (c - b) = 0.7738$
$\log \tan \frac{1}{2}(C + B) = 0.3586$

Adding $\qquad\qquad\qquad\qquad\quad 1.1324.$

$\log (c + b) = 1.4567$
$\log \tan \frac{1}{2}(C - B) = 9.6757$
$\frac{1}{2}(C - B) = 25° 22'$
$\frac{1}{2}(C + B) = 66° 21'$

Adding $\qquad\qquad\qquad\quad C = 91° 43'.$
Subtracting $\qquad\qquad\quad B = 40° 59'.$

To find the side a, we use the law of sines
$$a = \frac{b \sin A}{\sin B}.$$

$$\log b = 1.0546$$
$$\log \sin A = 9.8663$$
$$\overline{0.9209}$$
$$\log \sin B = 9.8168$$
$$\overline{\log a = 1.1041}$$
$$a = 12.71$$

Exercises and Problems

1. Given $a = 7, b = 9, C = 48°$, find c.
2. Given $a = 12, c = 10, B = 53° 17'$, find b.
3. Given $b = 100, c = 200, A = 29° 34'$, find a.
4. Given $a = 7, b = 9, C = 123° 20'$, find c.

Solve the following triangles, given:

5. $a = 147.1, b = 175.9, C = 43° 43'$.
6. $a = 147.1, b = 175.9, C = 137° 36'$.
7. $b = 9641, c = 8999, A = 67° 21'$.
8. $a = 0.3146, c = 0.8361, B = 83° 10'$.
9. $b = 25.25, c = 97.46, A = 98° 49'$.
10. $a = 1000, b = 100, C = 100°$.
11. Two automobiles start at the same time from the intersection of two roads which intersect at an angle of $76° 30'$. One automobile is traveling at the rate of 60 miles per hour, the other at 45 miles per hour. How far apart are they at the end of 15 minutes?
12. To find the distance between two points A and B separated by a small lake, a station C was chosen and the distances $CA = 4827$ yards, $CB = 3943$ yards, together with the angle $ACB = 67° 18'$ were measured. Find the distance from A to B.

ASSIGNMENT 19

Solution of Triangles (Areas)

43. Areas of triangles.

The area of a triangle is given in geometry as one half the product of the base by the altitude. With this as a starting point we can find expressions for the area in various forms. From Fig. 49 we have

$$\text{area} = K = \tfrac{1}{2}bh,$$

but $h = c \sin A$, hence

$$K = \tfrac{1}{2}bc \sin A. \quad (1)$$

In a similar manner we find

$$K = \tfrac{1}{2}ab \sin C, \quad (2)$$
$$K = \tfrac{1}{2}ac \sin B, \quad (3)$$

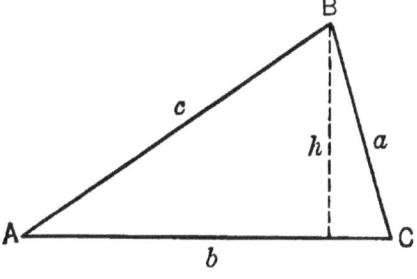

FIG. 49

giving the area in terms of two sides and the included angle.

Taking one of the above formulas, say

$$K = \tfrac{1}{2}ac \sin B$$

and replacing c by its value from the law of sines, $c = \dfrac{a \sin C}{\sin A}$, we have

$$K = \frac{a^2 \sin B \sin C}{2 \sin A} = \frac{a^2 \sin B \sin C}{2 \sin (B + C)}, \quad (4)$$

remembering $\sin A = \sin [180° - (B + C)] = \sin (B + C)$. Other expressions are

$$K = \frac{b^2 \sin A \sin C}{2 \sin (A + C)}, \quad (5)$$

$$K = \frac{c^2 \sin A \sin B}{2 \sin (A + B)}. \tag{6}$$

Starting again with the formula
$$K = \tfrac{1}{2}ac \sin B$$
and squaring, we have
$$K^2 = \frac{a^2c^2 \sin^2 B}{4} = \frac{a^2c^2(1 - \cos^2 B)}{4}$$
$$= \frac{ac}{2}(1 + \cos B)\frac{ac}{2}(1 - \cos B).$$

From the law of cosines, we have
$$K^2 = \frac{ac}{2}\left(1 + \frac{a^2 + c^2 - b^2}{2ac}\right)\frac{ac}{2}\left(1 - \frac{a^2 + c^2 - b^2}{2ac}\right),$$
$$= \frac{2ac + a^2 + c^2 - b^2}{4} \cdot \frac{2ac - a^2 - c^2 + b^2}{4}$$
$$= \frac{(a + c)^2 - b^2}{4} \cdot \frac{b^2 - (a - c)^2}{4}$$
$$= \frac{a + c - b}{2} \cdot \frac{a + c + b}{2} \cdot \frac{b - a + c}{2} \cdot \frac{b + a - c}{2}.$$

If we let $s = \dfrac{a + b + c}{2}$, we have
$$K^2 = s(s - a)(s - b)(s - c)$$
or
$$K = \sqrt{s(s - a)(s - b)(s - c)}. \tag{7}$$

44. Radius of inscribed circle.

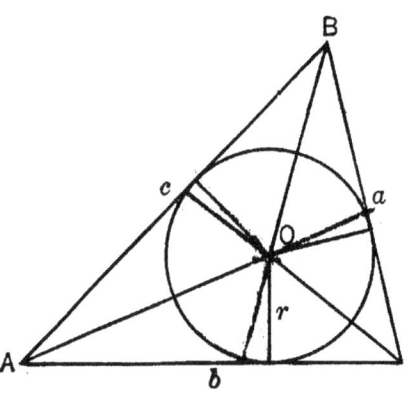

FIG. 50

From Fig. 50 we see that the area of the triangle ABC is equal to the sum of the areas of the triangles AOC, COB and AOB. The areas of these triangles are $\dfrac{br}{2}, \dfrac{ar}{2}, \dfrac{cr}{2}$, respectively, whence we have
$$K = rs. \tag{8}$$

RADIUS OF INSCRIBED CIRCLE

From equation (7) in Art. 43 we then find for the radius of the inscribed circle

$$r = \sqrt{\frac{(s-a)(s-b)(s-c)}{s}}. \qquad (9)$$

Exercises and Problems

Find the areas of the following triangles, using the parts given:

1. $b = 64.25$, $A = 24°\,23'$, $B = 61°\,48'$.
2. $a = 259.4$, $A = 47°\,10'$, $C = 59°\,17'$.
3. $a = 549.2$, $c = 835$, $A = 41°\,8'$.
4. $a = 840$, $b = 835$, $A = 41°\,8'$.
5. $A = 47°$, $b = 11$, $c = 17$.
6. $B = 47°\,18'$, $a = 11.34$, $c = 17.18$.
7. $a = 1653$, $b = 1777$, $c = 2131$.
8. $a = 2.136$, $b = 7.531$, $c = 6.714$.
9. Prove that the area of a triangle in terms of the side c and the angles A, B, C is

    $$\frac{1}{2} \frac{c^2 \sin A \sin B}{\sin C}.$$

10. Prove that the area of a parallelogram is equal to the product of two adjacent sides by the sine of the included angle.

ASSIGNMENT 20

Solution of Triangles (Case IV)

45. Case IV. Given three sides.

From theorems in elementary geometry, we know that in Fig. 51,

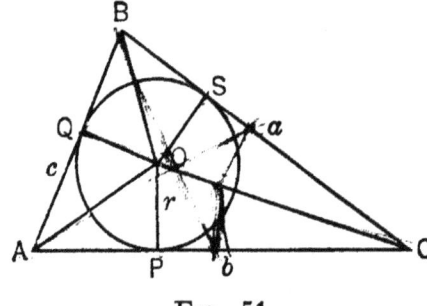

FIG. 51

$$AP = AQ,$$
$$BQ = BS,$$
$$CP = CS,$$

and that the angles A, B, C are bisected by the lines AO, BO, and CO respectively. Recalling that $2s = a + b + c$ we have

$$2s = BS + SC + AP + PC + AQ + QB$$
$$= 2BS + 2SC + 2AP$$

or $\quad s = BS + SC + AP = a + AP,$

whence $\quad AP = s - a. \quad (1)$

The line AO bisects the angle A. From the right triangle AOP we have

$$\tan \frac{A}{2} = \frac{r}{s-a}. \quad (2)$$

In a similar manner, we find

$$\tan \frac{B}{2} = \frac{r}{s-b}, \quad (3)$$

$$\tan \frac{C}{2} = \frac{r}{s-c}. \quad (4)$$

CASE IV. GIVEN THREE SIDES

EXAMPLE. Find the angles of the triangle whose sides are $a = 99$, $b = 100$, $c = 101$.

SOLUTION: First find $\log r$ where $r^2 = \dfrac{(s-a)(s-b)(s-c)}{s}$.

$2s = a + b + c = 300,$
$s = 150$
$s - a = 51$
$s - b = 50$
$s - c = 49$

$\log (s - a) = 1.7076$
$\log (s - b) = 1.6990$
$\log (s - c) = 1.6902$
adding $\quad\overline{5.0968}$
$\log s = 2.1761$
$\log r^2 = \overline{2.9207}$
$\log r = 1.4604$

Using the formulas

$$\tan\tfrac{1}{2}A = \frac{r}{s-a}, \qquad \tan\tfrac{1}{2}B = \frac{r}{s-b}, \qquad \tan\tfrac{1}{2}C = \frac{r}{s-c},$$

$\log r = 1.4604$	$\log r = 1.4604$	$\log r = 1.4604$
$\log s - a = 1.7076$	$\log s - b = 1.6990$	$\log s - c = 1.6902$
$\log \tan \tfrac{1}{2}A = \overline{9.7528}$	$\log \tan \tfrac{1}{2}B = \overline{9.7614}$	$\log \tan \tfrac{1}{2}C = \overline{9.7702}$
$\tfrac{1}{2}A = 29°\ 31'$	$\tfrac{1}{2}B = 30°\ 0'$	$\tfrac{1}{2}C = 30°\ 30'$
$A = 59°\ 2'$	$B = 60°\ 0'$	$C = 61°\ 0'$

CHECK. $A + B + C = 180°\ 2'$. It often happens that the sum of the computed angles differs from 180° due to doubling the unavoidable errors in finding the half angles. In finding $\log r$ we had the choice $\log r = 1.4604$, or $\log r = 1.4603$. We followed the computer's rule (Art. 32, footnote) and used the value ending in an even number. If we had used the other value, we would have found that $A + B + C = 180°$. If higher-place tables are used, we can of course find values of the angles which in general will give a better check.

Exercises and Problems

Find the angles of the following triangles to the nearest minute. Find the areas.

1. $a = 98, b = 100, c = 102$
2. $a = 24.13, b = 27.41, c = 31.31$
3. $a = 17.45, b = 20.00, c = 13.93$
4. $a = 2, b = 3, c = 4$
5. $a = 4, b = 5, c = 6$
6. $a = 127.6, b = 139.4, c = 21.63$
7. The diagonals of a parallelogram are 16.32 feet and 27.52 feet long. One side is 10 feet long. What are the angles of the parallelogram?
8. A triangular city lot has a frontage of 281.7 feet on one street and 256.3 feet on the other. The third side is 363.5 feet. Find the angle between the streets.

76 SOLUTION OF TRIANGLES (CASE IV)

9. Three points A, B, C are so located that $AB = 200$ yards, $AC = 126.6$, $BC = 213.6$ yards. B is due north of A. In what direction is C from A?

10. Two mountain tops are 27,140 yards apart. An observer is 18,140 yards from one and 28,370 yards from the other. What angle does the surveyor observe between the lines to the two mountains?

SUPPLEMENTARY EXERCISES AND PROBLEMS

Assignment 1

Construct the following angles:

1. 210°
2. − 330°
3. 278°
4. − 409°
5. − 285°
6. 315°
7. − 135°
8. 110°
9. − 240°

10. What positive angles are coterminal with the negative angles in Exercises 2, 4, 5, 7, 9?
11. What negative angles are coterminal with the positive angles in Exercises 1, 3, 6, 8?
12. Express each of the following angles in degrees: $\frac{\pi}{10}$ radians; $-\frac{3}{2}\pi$ radians; $\frac{2\pi}{3}$ radians; $\frac{\pi}{4}$ radians; $\frac{\pi}{6}$ radians.
13. A circle has a radius of 12 inches and an angle at its center subtends an arc of 4π feet in length. Find the angle in radians.
14. Assuming that the radius of the earth is 3960 miles, find how far a point on the equator travels in 1 hour; in 1 minute; in one second.

Assignment 2

1. Given $\cos A = \frac{12}{13}$, find the value of the remaining functions of A.
2. Given $\tan A = \frac{1}{7}$ and that a side $a = 3$. Construct the right triangle and write the values of the other functions.
3. Write all the functions for the angles A and B of a right triangle for which $c = 6$, $a = 4$.

EXERCISES AND PROBLEMS

In each of the following, write the trigonometric functions for both angle A and angle B.

4. Given $a = 8$, $b = 5$
5. Given $c = 11$, $a = 9$
6. Given $b = k$, $c = 1$
7. Given $\sin A = \frac{9}{15}$
8. Compare the values of the trigonometric functions of angle A in a triangle whose sides are $a = 3$, $b = 4$, $c = 5$, with those of a triangle with sides $a = 6$, $b = 8$, $c = 10$.
9. In a right triangle $a = \frac{1}{2}c$. Find the trigonometric functions of A.
10. Find the value of the product $(\cos A \cdot \tan A \cdot \sin B)$ if $\sec A = \sqrt{2}$.

Assignment 3

1. Can the sine and cosine of an angle ever have the same value? If so for what angles?
2. How do you account for the fact that $\sin 60°$ is not equal to twice $\sin 30°$?
3. If $\csc A = 2$, what is $\tan A$? What is the angle A?

By actual substitution, verify each of the following:

4. $\sin 30° \sec 30° = \tan 30°$.
5. $(\cot 60°)^2 - (\cos 60°)^2 = (\cot 60° \cdot \cos 60°)^2$.
6. $\cos 45° + \tan 45° \sin 45° = \sec 45°$.
7. $\sqrt{\dfrac{1 - \cos 60°}{1 + \cos 60°}} = \csc 60° - \cot 60°$.
8. Write each of the following as functions of the complementary angle: $\tan 38°$, $\cos 75°$, $\sin 19°$, $\sec 13°$, $\cot 45°$, $\csc 60°$.
9. If $\cot (60° - x) = \tan (15° + x)$, find the angle x.
10. If $\csc (28° + x) = \sec (14° - x)$, find the angle x.

EXERCISES AND PROBLEMS

Assignment 4

1. Show each of the following to be true:
$$\sin A = \sqrt{1 - \cos^2 A}, \quad \sin A = \frac{\tan A}{\sqrt{1 + \tan^2 A}}.$$

2. Express $\cos A$ in terms of each of the other functions.

3. Show that $\dfrac{\cos^2 \theta}{1 - \sin \theta} = 1 + \sin \theta$.

4. Are the identities given in Assignment 4 sufficient to enable us to express every function in terms of every other function?

Establish each of the following identities:

5. $(1 + \cos \theta)(\csc \theta - \cot \theta) = \sin \theta$.
6. $\csc^2 \theta \sec^2 \theta = \csc^2 \theta + \sec^2 \theta$.
7. $\cos^4 \theta - \sin^4 \theta = 1 - 2 \sin^2 \theta$.
8. $\dfrac{\cot^2 \theta - 1}{1 + \cot^2 \theta} = \cos^2 \theta - \sin^2 \theta$.
9. $(1 + \cot^2 x) \sin^2 x = 1$.
10. $\dfrac{\sin^2 x}{\tan^2 x} = 1 - \sin^2 x$.

Assignment 5

Compute the remaining parts and the area for the following right triangles, if given:

1. $a = 23, b = 18.1$.
2. $b = 13, c = 21$.
3. $A = 37° 28', a = 123$.
4. $c = 19.2, B = 51$.
5. $B = 38° 7', a = 35.6$.
6. In Fig. 52 a man M on the sixth floor of a building observes that the angle of elevation of the top T of another building 500 feet away is 53° 15', and the angle of depression of the bottom B of the building is 21° 35'. How tall is the building?

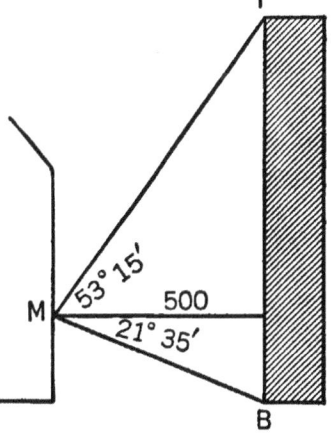

Fig. 52

EXERCISES AND PROBLEMS

7. In an oblique triangle (Fig. 53) given
$$A = 52°, B = 106°, BC = 13.8.$$
Find the other parts of the triangle and the area.

8. According to physics, when light passes from one medium to another, $\sin I = r \sin R$, where R is the angle of refraction, I is the angle of incidence, and r is the index of refraction. If $I = 31°$ and $r = 0.79$, find R.

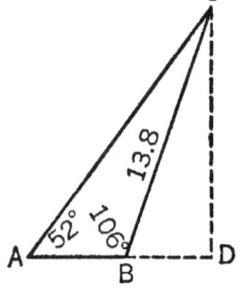

Fig. 53

9. Find the angle between the diagonal of a face of a cube and a diagonal of the cube.

10. Find the area of the parallelogram in Fig. 54.

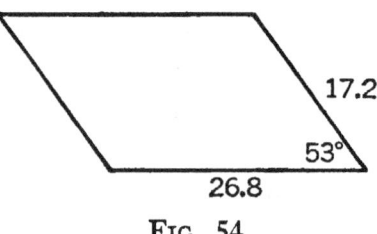

Fig. 54

Assignment 6

1. Join each of the following points to the origin and give the values of the six trigonometric functions for the positive angle thus formed.
$$P_1(-3, 4), P_2(12, -9), P_3(-2\sqrt{2}, -1), P(-7, -9).$$

2. If $\sin \theta = -\frac{2}{5}$ and $\cos \theta$ is positive, write all the other functions of θ.

3. If $\sin \theta = \frac{3}{5}$, how many angles are possible which are less than 360°? What are the approximate values of the angles? How do the functions differ for these angles?

4. Draw the angles less than 360° for which $\tan \theta = -\frac{1}{4}$ and give the values of the other functions for these angles.

5. Write the polar coordinates for each of the points given in Exercise 1.

6. How many ratios must be given to fix definitely the terminal side of an angle? Can these ratios be given arbitrarily? (For example,

EXERCISES AND PROBLEMS

does $\sin \theta = \frac{2}{5}$, $\csc \theta = \frac{5}{2}$ determine definitely the position of the terminal side of the angle θ?)

7. Sketch the angle for which $\tan \theta = -\frac{3}{4}$ and $\sin \theta$ is positive.
8. State for each quadrant the functions which are negative.

Assignment 7

By means of the rule in Art. 16, write equivalent expressions in terms of acute angles for:

1. $\sin 109°$
2. $\cos 255°$
3. $\tan 263°$
4. $\cot 328°$
5. $\cos (-105°)$
6. $\sec (-400°)$
7. $\csc 242°$
8. $\tan (-113°)$
9. $\sin (-118°)$

10. Using the rule for cofunctions of complementary angles, write the results you obtained in Exercises 1 to 9 in terms of an angle which is less than 45°. (See Art. 5.) Thus:
$$\sin 109° = \sin (180° - 71°) = \sin 71° = \cos 19°.$$

Assignment 8

Draw the graphs of each of the following:

1. $y = 2 \sin x$
2. $y = 3 \cos x$
3. $y = \sin 2x$
4. $y = 2 \sin 2x$

5. $y = \sin x + \cos x$. *Hint:* From the graphs of $y = \sin x$ and $y = \cos x$, draw a graph whose ordinates corresponding to each value of x are the algebraic sum of the ordinates of $y = \sin x$ and $y = \cos x$.

6. $y = \sin x - \cos x$
7. $y = \sin \frac{x}{2}$
8. $y = \frac{1}{2} \sin x$

Solve for angles less than 360° which satisfy the following equations:

9. $2 \sin x (1 - 2 \cos x) = 0$
10. $\sin^2 x = 2 \cos^2 x$
11. $\sin x + \cos x = \sqrt{2}$
12. $2 \sin^2 x + 3 \cos x = 0$

Assignment 9

1. If $\sin x = \frac{3}{4}$ and $\cos y = -\frac{4}{5}$, x and y in the second quadrant, find the values of $\sin(x+y)$, $\cos(x+y)$, $\tan(x+y)$.
2. With the same angles x and y as in Exercise 1, find $\sin(x-y)$, $\cos(x-y)$, $\tan(x-y)$.
3. Show that $\tan\left(\dfrac{\pi}{6} - y\right) = \dfrac{\sqrt{3} - \tan y}{1 + \sqrt{3}\,\tan y}$.
4. Find $\tan(x+y)$ and $\tan(x-y)$ if $\sin x = \frac{3}{5}$, $\sin y = \frac{5}{13}$, and x and y are acute angles.
5. Using the method of Art. 24, derive a formula for $\cot(x+y)$ in terms of $\cot x$ and $\cot y$.
6. Using the method of Art. 24, derive a formula for $\cot(x-y)$ in terms of $\cot x$ and $\cot y$.
7. Prove the identity
$$\frac{\tan(x+y)}{\cot(x-y)} = \frac{\tan^2 x - \tan^2 y}{1 - \tan^2 x \tan^2 y}.$$
8. Show that $\sin(x+y)\cos y - \cos(x+y)\sin y = \sin x$.
9. Write as a single trigonometric function of an angle,
$$\sin 36° \cos 24° + \cos 36° \sin 24°.$$
10. Using the addition formulas, compute the following: sine, cosine and tangent of 105°; sine, cosine and tangent of 210°.

Assignment 10

1. Find the values of sine and cosine of 150°, using function of 75°.
2. Find the sine, cosine and tangent of 90°, using functions of 45°.
3. If $\csc x = -\frac{13}{12}$, and x is in the third quadrant, find $\cos 2x$.
4. If $\tan x = -\frac{3}{4}$ and x is in the second quadrant, find the value of $\sin 2x$.
5. Show that $\dfrac{2\tan x}{1 + \tan^2 x} = \sin 2x$.
6. Find the values of the sine, cosine and tangent of 105°, using functions of 210°.

7. If $\tan x = -\frac{5}{12}$, and x is in the second quadrant, find the values of the sine, cosine and tangent of $\frac{x}{2}$.

8. Express, in terms of radicals, the values of $\sin \frac{\pi}{4}$, $\cos \frac{\pi}{4}$, $\tan \frac{\pi}{4}$.

9. From $\cos 270° = 0$, find $\tan 135°$ and $\cot 135°$.

10. Show that $\sec \frac{\theta}{2} = \frac{1}{\cos \frac{\theta}{2}} = \pm \sqrt{\frac{2 \sec \theta}{1 + \sec \theta}}$.

Assignment 11

1. Show that $\sin 75° - \sin 15° = \cos 45°$.
2. Show that $\cos 75° + \cos 45° = \cos 15°$.
3. Express $2 \cos 8\theta \cdot \sin 2\theta$ as the difference of two sines.
4. Simplify the expression $\frac{\sin 80° - \sin 60°}{\cos 80° + \cos 60°}$.
5. Express $\cos x + \sin 2x$ as a product.
6. Show that $\frac{\sin 2A + \sin 2B}{\cos 2A + \cos 2B} = \tan (A + B)$.
7. Show that $\frac{\sin 3x - \sin x}{\cos 3x + \cos x} = \tan x$.
8. Show that $\frac{\cos 2x - \cos 3x}{\sin 2x + \sin 3x} = \tan \frac{x}{2}$.
9. Give a single numerical value for $\frac{\sin 73° + \sin 47°}{\cos 73° + \cos 47°}$.
10. Express $2 \sin x \cos 7x$ as a difference of two functions.

Assignment 12

1. Change each of the following statements to the exponential form:
   (a) log 12 = 0.0792
   (b) log 1 = 0
   (c) log 10 = 1
   (d) log 0.0001 = −4
2. Change each of the following to the logarithmic form:
   (a) $10^6 = 1000000$
   (b) $10^1 = 10$
   (c) $10^{-3} = 0.001$
   (d) $\sqrt{10} = 3.162$
3. Prove log 1 = 0.
4. Prove log 10 = 1.
5. Solve the following equations for x:
   (a) log x = 3
   (b) log x = 0.3010
   (c) log x = 0.8451
   (d) log 100 = x
   (e) log 0.1 = x
   (f) log 20 = x
6. Given log 20 = 1.3010, log 300 = 2.4771, log $\sqrt{5}$ = 0.3495, find:
   (a) log 2
   (b) log 3
   (c) log 5
   (d) log 2000
   (e) log 3000
   (f) log 2700

Express the logarithms of the following in terms of the logarithms of prime numbers.

7. $\sqrt{6.6}$.
8. $\sqrt[3]{1.87}$.
9. 23100.
10. $\sqrt[5]{2.5}$.

Assignment 13

1. Find log 7163.
2. Find log 8.888.
3. Find log 0.7189.
4. Find log 0.009376.
5. Find the logarithms of the following numbers: 28431, 28432, 28433, 28434, 28437.
6. What are the characteristics of the logarithms of the following numbers: 6193, 81563, 1.875, 500000, 0.8378, 0.001896?
7. Log x = 2.9176. Find x.
8. Log y = 9.7184 − 10. Find y.

EXERCISES AND PROBLEMS

9. Log $z = 0.1211$. Find z.
10. Where is the decimal point in the numbers corresponding to the following logarithms: 0.7183, 9.7142 − 10, 4.2718, 7.1234 − 10, 1.7298?

Assignment 14

Compute to four significant figures by the use of logarithms.

1. $\dfrac{1}{(7.832)^3}$.
2. $\dfrac{\sqrt{2}}{\sqrt[3]{3}}$.
3. $\sqrt[10]{10}$.
4. $\dfrac{7\pi}{827}$.

5. Solve for x if $x^2 = \log(84.73)^3$.
6. Find the volume and surface of a sphere of radius 7.136.
7. $1000 is put at interest at 4 per cent compounded annually. What will it amount to at the end of 10 years? At the end of 20 years? (See Example 5, Art. 34.)
8. A meter is 39.37 inches. What is the length in meters of a track 100 yards long?
9. The area of a triangle whose sides are a, b, and c is given by the formula,
$$\text{Area} = \sqrt{s(s-a)(s-b)(s-c)} \quad \text{where} \quad s = \frac{a+b+c}{2}.$$

   Find the area of the triangle whose sides are 7.182, 8.182, 9.182.
10. What is the weight in tons of an iron sphere whose radius is 2.236 feet, if the weight of a cubic foot of water is 1000 ounces and the specific gravity of the iron is 7.154?

Assignment 15

1. Given $c = 1000$, $a = 100$, solve the right triangle.
2. Given $b = 1000$, $a = 100$, solve the right triangle.
3. Given $A = 1.2$ radians, $c = 193.1$, solve the right triangle.

EXERCISES AND PROBLEMS

4. Find A and B in radian measure, for the right triangle, given $a = 100$, $b = 101$.

5. At what angle does a road incline to the horizontal if the grade is 15 in 100 (that is, a rise of 15 feet in 100 feet measured horizontally)?

6. From the top of a mountain 2070 feet above a level plain, the angles of depression of two points in the same vertical plane as the observers are 37° 6′ and 42° 25′. How far apart are the points?

7. In a right triangle the sides a and b are 10 and 20 inches respectively. What change is made in the angle A if the side b is lengthened to 21 inches?

8. In the triangle of Problem 7, what change is made in the angle A if the side a is lengthened to 11 inches, the side b remaining at 20 inches?

9. In the triangle of Problem 7, what change is made in the angle A if both a and b are increased 1 inch in length?

10. The town A is 21.75 miles directly north of the town B. The town C is 27.50 miles directly east of B. What is the direction of the road from C to A?

Assignment 16

1. Given $a = 1200$, $A = 67°$, $C = 53°$, find b, c, B.
2. Given $b = 1476$, $B = 50°\ 51′$, $C = 27°\ 49′$, find a, c, A.
3. Given $c = 0.7631$, $A = 100°$, $B = 40°$, find a, b, C.
4. Let d be the diagonal of a parallelogram and A and B the angles which the diagonal makes with the sides. Find the sides of the parallelogram when $d = 12.36$, $A = 19°\ 36′$, $B = 43°\ 6′$.
5. Let $ABCD$ be a trapezoid for which AB and DC are parallel sides. Find the nonparallel sides when $AB = 15.26$, $DC = 7.13$, $DAB = 69°\ 35′$, $CBA = 40°\ 40′$.
6. The angles of a triangle are in the proportion 6 : 11 : 23, and the side opposite the smallest angle is 25.3 inches. Find the angles and the other sides.

EXERCISES AND PROBLEMS

7. Given $A = 6° 30'$, $C = 5° 41'$, $c = 10$, find a, b, B.
8. Two angles of a triangle are $A = 40°$ and $B = 80°$. Find the ratio $\frac{a}{b}$ of the opposite sides.
9. Two observers 7 miles apart, facing one another on a long level road, find that the angles of elevation of a balloon in the same vertical plane with themselves are 66° and 56°. Find the height of the balloon.
10. What effect does doubling the angles A and C in Problem 7 have on the sides a and b?

Assignment 17

1. Given $a = 1736$, $b = 2845$, $B = 143°$, find A, C, c.
2. Given $b = 2013$, $c = 2012$, $C = 103° 17'$, find a, A, B.
3. Given $a = 2329$, $c = 1417$, $C = 37° 59'$, find b, A, B.
4. Given $c = 5366$, $b = 5314$, $B = 81° 55'$, find a, A, C.
5. Given $a = 11$, $c = 10$, $A = 47° 13'$, find b, B, C.
6. Given $a = 12$, $c = 11$, $A = 47° 13'$, find b, B, C.
7. Given $b = 1000$, $c = 2000$, $C = 165°$, find a, A, B.
8. The shorter side of a parallelogram is 25.37. The longer diagonal is 66.78. The acute angle between diagonals is 41° 37'. Find the other diagonal.
9. A tree 86 feet high standing on a slope casts a shadow 146 feet long down the slope. At the end of the shadow the angle subtended by the tree is 30° 5'. What is the elevation of the sun?

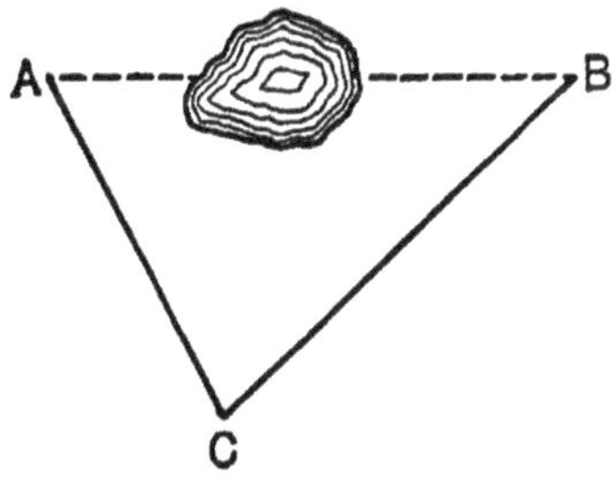

FIG. 55

10. A and B are two points on opposite sides of a lake. It is desired to find the distance AB. A point C is chosen, easily accessible from A and B. It is found that $AC = 7738$ feet and $CB = 8147$ feet. At A the angle BAC is found to be 61° 17'. Find AB.

Assignment 18

1. Given $a = 10$, $b = 25$, $C = 47° 39'$, find c.
2. Given $b = 1$, $c = 3$, $A = 103° 40'$, find a.
3. Given $a = 2713$, $c = 3941$, $B = 149° 31'$. Find b, A, C.
4. Given $a = 2713$, $c = 3941$, $B = 148° 31'$. Find b, A, C.
5. Given $b = 1000$, $c = 4000$, $A = 10°$, find a, B, C.
6. Given $b = 2000$, $c = 8000$, $A = 20°$, find a, B, C.
7. Given $a = 3400$, $b = 340$, $C = 171° 45'$, find c, A, B.
8. Given $b = 7.183$, $c = 9.246$, $A = 45°$, find a, B, C.
9. A parallelogram has a side 10 inches long and a diagonal 12 inches long. The angle between the diagonal and the side is $37° 15'$. Find the other side and other diagonal.
10. At a signal station two explosions are seen and heard from two directions. The angle between the two directions is $49° 30'$. The interval between the flash and sound of the explosion for one was $4\frac{1}{2}$ seconds, for the other $6\frac{1}{2}$ seconds. How far apart are the two explosions? (Take 1140 feet per second as the velocity of sound.)

Assignment 19

Find the areas of the following triangles:

1. $a = 201.3$, $A = 21° 23'$, $B = 102° 16'$.
2. $b = 201.3$, $A = 21° 23'$, $B = 102° 16'$.
3. $c = 201.3$, $A = 21° 23'$, $B = 102° 16'$.
4. $b = 13.61$, $c = 21.47$, $A = 57° 13'$.
5. $a = 2000$, $c = 100$, $B = 162° 30'$.
6. $a = 341.2$, $b = 536.7$, $c = 716.1$.
7. Prove that the area of any quadrilateral is equal to half the product of the diagonals by the sine of the included angle.
8. Find the difference between the areas of the two triangles given by $a = 625$, $c = 835$, $A = 41° 8'$.

EXERCISES AND PROBLEMS

9. Find an expression for the area of an isosceles trapezoid in terms of the two parallel sides and an acute angle.
10. How many acres are there in a triangular field with sides 1470, 1638, 1259 feet long? (One acre is 43,560 square feet.)

Assignment 20

Find the angles and areas of the following triangles:

1. $a = 9, b = 10, c = 11$.
2. $a = 479, b = 579, c = 679$.
3. $a = 1236, b = 1589, c = 1339$.
4. $a = 7148, b = 8376, c = 3150$.
5. $a = 300, b = 400, c = 500$.
6. Using the law of cosines, find the smallest angle in the triangle whose sides are 8, 10, 12.
7. What is the radius of the largest circle that can be drawn in the triangle whose sides are 3, 4, 5 feet long?
8. A surveyor's chain is 66 feet long. The sides of a triangular field are 14.6, 18.7, 23.6 chains. What is the area in square feet?
9. Two sides of a triangular field are 163.7 and 194.6 chains. The area of the field is 1590 acres. What is the third side? (Ten square chains = 1 acre.)
10. The sides of a triangle are 17, 21, 18. A line is drawn from the midpoint of the longest side to the vertex opposite. Find the length of the line.

MISCELLANEOUS EXERCISES

1. Give positive angles which are coterminal with each of the following: $-118°, -321°, -154°, -315°, -210°, -53°$.
2. Show that the area of a segment of a circle (shaded portion, Fig. 56) equals $\frac{1}{2}r^2(\theta - \sin\theta)$.
3. Express each of the following angles in degrees:

   $$\frac{\pi}{5}, \frac{\pi}{7}, \frac{2\pi}{9}, \frac{\pi}{18}, \frac{11\pi}{18}, \frac{\pi}{15}, \frac{7\pi}{20}, \frac{\pi}{30}.$$

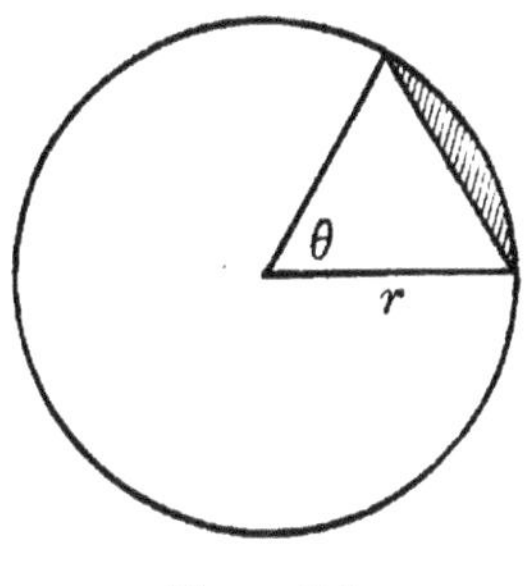

4. A railway tank car is 38 feet long and has flat circular ends which are 8 feet in diameter. If a measuring stick shows that the depth of the gasoline in the car is 2.5 feet, how many gallons are in the car? (Use 1 cu. ft. = 7.5 gallons, and formula of Problem 2.)

   FIG. 56

5. Assuming that the earth travels in a circular orbit, around the sun, of radius 93,000,000 miles, find the length of arc through which it moves in one second. (Use 365 days = 1 revolution.)
6. Find all the functions of angles A and B in a right triangle if $c = 17$ and $a = 5$.
7. Give the values of the functions of A if $\csc A = -\frac{17}{8}$.
8. Construct a right triangle for which $\tan A = \frac{8}{15}$ and $c = 34$.
9. Verify $\dfrac{\sec A - 1}{\sec A + 1} = \dfrac{1 - \cos A}{1 + \cos A}$ for $A = 60°; 45°; 30°$.
10. Verify $\dfrac{1 + \tan A}{\sec A} = \sin A$ for $A = 60°; 45°; 30°$.
11. In a right triangle, let the side opposite angle A be designated by $\sin A$ and let the hypotenuse be 1. Show, by using Fig. 57, that forming the ratios of the sides will give the functions of angle A in terms of $\sin A$.

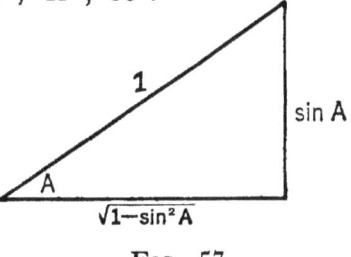

FIG. 57

MISCELLANEOUS EXERCISES

12. As in Exercise 11, let the hypotenuse be sec A and the horizontal side be 1. The opposite side is then $\sqrt{\sec^2 A - 1}$. Find all the functions of angle A in terms of the secant.

13. By choosing appropriate triangles in Fig. 24 of Art. 15, show that one may obtain the identities (III) to (VII) from the figure.

14. An angle at the center of a circle of radius 3.6 in. subtends an arc equal to 13.5 inches. Find the central angle.

15. Show that $\sin\left(\dfrac{\pi}{3} + \theta\right) - \sin\theta = \sin\left(\dfrac{\pi}{3} - \theta\right)$.

16. Show that $\tan\left(\dfrac{\pi}{2} + \dfrac{x}{2}\right) = \sec x + \tan x$.

17. Show that $\dfrac{\sin^2 x - \sin^2 y}{\sin x \cos x - \sin y \cos y} = \tan(x+y)$.

18. Show that $\sin 2x \sec x = 2 \sin x$.

19. Show that $\csc 2x \cdot \cos x = \tfrac{1}{2} \csc x$.

20. Show that $\dfrac{\sin(A+B)}{\sin A \sin B} = \cot A + \cot B$.

21. Show that $\dfrac{\sin 4x}{\sin 2x} = 2 \cos 2x$.

22. Find the area between a circle of radius 4 in. and the inscribed regular pentagon.

23. Find the area between a circle of radius 4 in. and the circumscribed regular hexagon.

24. Find the difference between the perimeters of a regular pentagon inscribed and one circumscribed to a circle of radius 3 inches.

25. In order to find the distance between two points A and C on opposite sides of a river, CB was laid off at right angles to AC. $CB = 150$ feet; angle CBA was found to be 82° 13'. Find AC. (Fig. 58.)

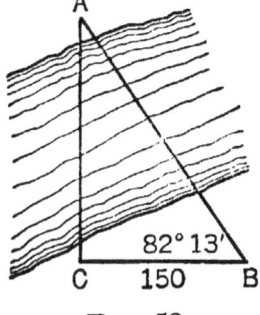

Fig. 58

26. We learn in physics, that when light passes from one medium to another, $\sin I = r \sin R$, where R is the angle of refraction, I is the angle of inci-

MISCELLANEOUS EXERCISES

dence, and r is the index of refraction. If $I = 32°$ and $r = .81$, find R.

27. Plot the points having the following polar coordinates and find their rectangular coordinates.

$$(5, 180°), \ (4, \pi), \ \left(4, \frac{4\pi}{3}\right), \ (0, 2), \ \left(-5, \frac{\pi}{2}\right).$$

28. Change the following rectangular coordinates to polar coordinates.

$$(3, 4), \ (-3, 4), \ (2, 8), \ (1, \sqrt{3}), \ (2\sqrt{3}, 2).$$

29. Draw the graph of the line $3x - 4y = 0$ and find the acute angle it makes with the positive x-axis. Give the six functions of this angle.

30. Draw the graph of the line $2x - 3y = 6$ and determine the acute angle which it makes with the positive x-axis. Write the sine of this angle.

31. Find the acute angle between the lines of Exercises 29 and 30.

Write an equivalent expression for each of the following in terms of an angle which is less than 45°.

32. sin 113°
33. cos 233°
34. tan 247°
35. sec − 118°
36. cot 320°
37. csc 418°
38. tan 310°
39. cos − 190°
40. sin 341°
41. cot 161°

Draw the graphs of the following equations:

42. $y = \tfrac{1}{2} \cos x$
43. $y = \cos 2x$
44. $y = 2 \cos x$

45. Compare the graphs in Exercises 42, 43, 44.

46. $y = x + \sin x$.
  Hint: Add ordinates of the graphs of $y = x$ and $y = \sin x$ to obtain y values for the graph.

47. Show that $\dfrac{\sin (x+y)}{\sin (x-y)} = \dfrac{\tan x + \tan y}{\tan x - \tan y}$.

48. Show that $\dfrac{\sin (x+y)}{\cos x \cos y} = \tan x + \tan y$.

49. Show that $\sin (x+y) \cdot \sin (x-y) = \sin^2 x - \sin^2 y$.

50. Show that $\tan \frac{x}{2} = \csc x - \cot x$.

Solve for angles, less than 360°, which satisfy the following equations.

51. $\csc x + \cot x = \sqrt{3}$.

52. $\tan^2 x + \cot^2 x = 2$.

53. $\sin^2 x - 2 \cos x + \frac{1}{4} = 0$.

54. $\sin 2x + \sin x = 0$.

55. $2 \sin^2 x - 3 \sin x + 1 = 0$.

56. In the United States a billion is one thousand millions. In England it is a million millions. What is the logarithm of a million according to each of these definitions?

57. Write as the logarithm of a single number:
$$\tfrac{1}{2} \log 4 + \tfrac{1}{3} \log 27 - \tfrac{1}{4} \log 16.$$

58. Express log 0.273 in terms of logarithms of prime numbers.

59. What is the characteristic of $\log (2736 \cdot 10^{-9})$?

60. Log $x = 7.0000$. What is x?

61. The geometric mean of n numbers is the nth root of their product. Find the geometric mean of 2.174, 3.174, 4.174 to four significant figures.

62. A square mile contains 640 acres. One kilometer equals 0.6214 miles. What part of a square kilometer is one acre?

63. Find $(0.1)^{0.1}$.

64. A ladder 42 feet long may be so placed that it will reach a window 33.5 feet high on one side of a passageway, and by turning it over without moving the foot, it will reach a window 22 feet high on the other side. How wide is the passageway?

65. Two lights with a range of 32 miles each are to be set up on a straight coast. How far apart can they be placed so that a ship sailing parallel to the coast and 25 miles offshore will not lose sight of one before the other comes into view?

66. Express the area of a right triangle in terms of a and A.

67. Show that in any triangle $c = a \cos B + b \cos A$.

68. One side of a parallelogram is 4.6 feet long and the angles which

MISCELLANEOUS EXERCISES

the diagonals make with this side are 35° 56′ and 52° 37′. Find the lengths of the diagonals.

69. A flagstaff, h feet high, stands on the edge of a cliff. From a point in the plain extending away from the foot of the cliff, the angles of elevation of the top and bottom of the flagstaff are observed to be A and B respectively. Show that the height of the cliff is
$$\frac{h \sin B \cos A}{\sin (A - B)}.$$

70. How many triangles are determined by the given parts $A = 60°$, $b = 1000$, $a = 500$; $a = 866$; $a = 900$; $a = 1000$; $a = 2000$?

71. The outer surface of the walls of a fort inclines so that the angle made with the horizontal is 92° 30′. A ladder 42 feet long, with its foot 16 feet from the base, leans against the wall. How far up the wall is the top of the ladder?

72. Show that in any triangle $\cos A = \dfrac{b^2 + c^2 - a^2}{2bc}$.

73. The diagonals of a parallelogram are 20.4 inches and 23.6 inches long. The acute angle between them is 47° 13′. What are the lengths of the sides of the parallelogram?

74. Two sides and the included angle of a triangle are measured and found to be 37.8 inches, 74.3 inches and 27° 44′ respectively. What error would be made in computing the third side if an error of two minutes was made in reading the angle?

75. Let R be the radius of the circle circumscribing the triangle ABC (Fig. 59). Prove that $R = \dfrac{b}{2 \sin B}$.

76. Prove that the area of a triangle is equal to $\dfrac{abc}{4R}$, where R is the radius of the circumscribing circle.

77. Show that the area of a regular pentagon of which one side is a is $\dfrac{5a^2}{4} \cot 36°$.

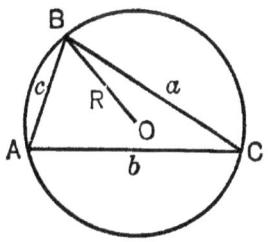

Fig. 59

78. Show that the area of a regular pentagon inscribed in a circle of radius r is $\tfrac{5}{2}r^2 \sin 72°$.

79. Using the expressions for $\sin \dfrac{A}{2}$ and $\cos \dfrac{A}{2}$ given in Art. 26, show that if A is an angle of a triangle whose sides are a, b, c, then

$$\sin \frac{A}{2} = \sqrt{\frac{(s-b)(s-c)}{bc}}, \quad \cos \frac{A}{2} = \sqrt{\frac{s(s-a)}{bc}}.$$

80. Find $\sin A$ in terms of a, b, c and s.

81. The centers of three circles of radius 10 lie on the vertices of a triangle of sides 25, 35, 40. What is the area of the smallest triangle which will enclose all three circles?

ANSWERS TO ODD-NUMBERED EXERCISES AND PROBLEMS

Answers to numerical exercises or problems were found in general by the use of four-place tables. Hence the fourth significant digit is doubtful and in some cases will vary with different methods of solution.

Assignment 1, Art. 2, page 6

1. $\frac{\pi}{6}, \frac{\pi}{4}, \frac{\pi}{3}, \frac{\pi}{2}, \frac{2\pi}{3}, \frac{3\pi}{4}, \frac{5\pi}{6}, \pi, \frac{7\pi}{6}, \frac{5\pi}{4}, \frac{3\pi}{2}, \frac{5\pi}{3}, \frac{7\pi}{4}, \frac{11\pi}{6}, 2\pi$.
3. 2911 feet.   5. 50,266 feet per minute.   7. $r = 2$ inches.

Assignment 2, Art. 3, pages 8, 9

1. Corresponding ratios are the same.
3. Sine and cosine. Secant and cosecant.
5. Sine and tangent small, nearly zero, cosine nearly 1. Sine nearly 1, tangent increasing beyond bound, cosine small, nearly zero.

Assignment 3, Art. 4, page 11

1. Same as the values given in Art. 4.   3. $\sqrt{3} = \sqrt{3}$.   5. $\frac{\sqrt{3}}{2} = \frac{\sqrt{3}}{2}$.

Art. 5, page 12

1. $\cos 27°$, $\sin 52°$, $\cot 19°$, $\tan 65°$, $\csc 75°$, $\sec 9°$.
3. $x = 30°$.   5. $A = 7° 30'$.

Assignment 5, Art. 9, page 19

1. 0.6106.   3. 0.7280.   5. $40° 45'$.

Art. 10, pages 21, 22

1. $A = 52° 36'$, $B = 37° 24'$, $c = 21.40$, area $= 110.5$.
3. $B = 30°$, $a = 42.09$, $c = 48.6$, area $= 511.4$.
5. $A = 71° 55'$, $a = 83.30$, $c = 87.63$, area $= 1133$.
7. 3.527 feet.   9. 2803 feet.
11. $A = B = 68° 20'$, $AD = 44.43$, altitude $= 41.28$.
13. 6575 feet.   15. 28.3 feet.

ANSWERS

Assignment 6, Art. 13, page 27

1.

	Sin	Cos	Tan	Cot	Sec	Csc
P_1	$\frac{4}{5}$	$\frac{3}{5}$	$\frac{4}{3}$	$\frac{3}{4}$	$\frac{5}{3}$	$\frac{5}{4}$
P_2	$\frac{5}{\sqrt{74}}$	$\frac{7}{\sqrt{74}}$	$\frac{5}{7}$	$\frac{7}{5}$	$\frac{\sqrt{74}}{7}$	$\frac{\sqrt{74}}{5}$
P_3	$\frac{3}{\sqrt{10}}$	$-\frac{1}{\sqrt{10}}$	-3	$-\frac{1}{3}$	$-\sqrt{10}$	$\frac{\sqrt{10}}{3}$
P_4	$-\frac{5}{\sqrt{34}}$	$-\frac{3}{\sqrt{34}}$	$\frac{5}{3}$	$\frac{3}{5}$	$-\frac{\sqrt{34}}{3}$	$-\frac{\sqrt{34}}{5}$
P_5	$-\frac{1}{\sqrt{5}}$	$\frac{2}{\sqrt{5}}$	$-\frac{1}{2}$	-2	$\frac{\sqrt{5}}{2}$	$-\sqrt{5}$

3.

	30°	45°	60°	120°	135°	150°	210°	225°	240°	300°	315°	330°
sin	$\frac{1}{2}$	$\frac{\sqrt{2}}{2}$	$\frac{\sqrt{3}}{2}$	$\frac{\sqrt{3}}{2}$	$\frac{\sqrt{2}}{2}$	$\frac{1}{2}$	$-\frac{1}{2}$	$-\frac{\sqrt{2}}{2}$	$-\frac{\sqrt{3}}{2}$	$-\frac{\sqrt{3}}{2}$	$-\frac{\sqrt{2}}{2}$	$-\frac{1}{2}$
cos	$\frac{\sqrt{3}}{2}$	$\frac{\sqrt{2}}{2}$	$\frac{1}{2}$	$-\frac{1}{2}$	$-\frac{\sqrt{2}}{2}$	$-\frac{\sqrt{3}}{2}$	$-\frac{\sqrt{3}}{2}$	$-\frac{\sqrt{2}}{2}$	$-\frac{1}{2}$	$\frac{1}{2}$	$\frac{\sqrt{2}}{2}$	$\frac{\sqrt{3}}{2}$
tan	$\frac{1}{\sqrt{3}}$	1	$\sqrt{3}$	$-\sqrt{3}$	-1	$-\frac{1}{\sqrt{3}}$	$\frac{1}{\sqrt{3}}$	1	$\sqrt{3}$	$-\sqrt{3}$	-1	$-\frac{1}{\sqrt{3}}$

5. Third, fourth. **7.** First.

Assignment 7, Art. 16, page 32

1. sin 17°. **3.** $-$ cot 42°. **5.** $-$ sec 71°. **7.** csc 15°. **9.** $-$ tan 43°.

Assignment 8, Art. 21, page 37

1. $x = 30°, 270°$. **3.** $x = 36° 52', 143° 8'$.
5. $x = 30°, 150°$. **7.** $x = 120°, 240°$. **9.** $x = 120°, 240°$.

Assignment 9, Art. 24, pages 40, 41

1. $\sin(x+y) = \dfrac{8 + 3\sqrt{5}}{15}$, $\cos(x+y) = \dfrac{4\sqrt{5} - 6}{15}$,
$\tan(x+y) = \dfrac{25\sqrt{5} + 54}{22}$.

ANSWERS

3. $\sin(x-y) = \dfrac{-11}{\sqrt{377}}$, $\cos(x-y) = \dfrac{16}{\sqrt{377}}$, $\tan(x-y) = -\dfrac{11}{16}$.

5. $\sin 75° = \sin(30° + 45°) = \tfrac{1}{4}(\sqrt{6} + \sqrt{2})$, $\cos 75° = \tfrac{1}{4}(\sqrt{6} - \sqrt{2})$, $\tan 75° = 2 + \sqrt{3}$.

7. $\dfrac{6 - 4\sqrt{5}}{15}$, $\dfrac{3\sqrt{5} - 8}{15}$.

Assignment 10, Art. 25, page 43

1. $\dfrac{\sqrt{3}}{2}, \dfrac{1}{2}, \sqrt{3}$.

Art. 26, pages 44, 45

1. $\dfrac{1}{2}, \dfrac{\sqrt{3}}{2}, \dfrac{1}{3}\sqrt{3}$.   7. $\tan \theta = \pm 3\tfrac{3}{7}$.

Assignment 11, Art. 27, page 47

1. $2 \sin 35° \cos 5°$.   3. $-2 \sin x \sin 4x$.
5. $\sin 8x + \sin 2x$.   7. $\cos 4\theta + \cos 2\theta$.

Assignment 12, Art. 28, page 49

1. (a) $\log 100000 = 5$.   (b) $\log 1700 = 3.2304$.
   (c) $\log 23 = 1.3617$.   (d) $\log 0.1 = -1 = 9.0000 - 10$.
   (e) $\log 0.743 = -0.1290 = 9.8710 - 10$.

Art. 29, page 50

1. (a) 0.7781; (b) 1.6232; (c) 0.6990; (d) 1.6902; (e) 0.5441;
(f) 0.2386; (g) 0.2817; (h) 5.7304; (i) 5.2464; (j) 1.0212.

3. $3 \log 5 + \tfrac{1}{2} \log 7$.

5. $\tfrac{1}{6} \log 3 - \tfrac{3}{4} \log 2 - \tfrac{1}{4} \log 5$ or $\tfrac{1}{6} \log 3 - \tfrac{1}{2} \log 2 - \tfrac{1}{4}$.

7. $\log 2 + \log 3 + \log 5 + \log 7$.   9. $\tfrac{2}{3}(\log 2 + \log 5)$.

11. $\tfrac{1}{2} \log 5 - \tfrac{2}{3}(\log 2 + \log 3)$.

Assignment 13, Art. 31, page 53

1. 2.5328.   3. 0.9996.   5. $9.8722 - 10$.   7. 4.8549.   9. $6.7924 - 10$.

Art. 32, page 54

1. 3.4335.   3. 3.0000.   5. 0.5372.   7. 5.6824.   9. $8.9927 - 10$.

Art. 33, page 54

1. 2.441.   3. 88.52.   5. 6124.   7. 0.04774.   9. 1064.

Assignment 14, Art. 34, page 56

1. 37220.   3. 0.02672.   5. 0.8217.   7. 116.1.   9. 9705.
11. 1.030.   13. 0.2461.   15. 0.6934.   17. 0.6156.   19. 1.062.

Assignment 15, Art. 35, page 59

1. $a = 61.8, b = 102.9, B = 59°$.
3. $b = 7948, c = 7972, B = 85° 25'$.
5. $b = 2.221, c = 3.118, A = 44° 35'$.
7. $a = 13.69, c = 21.77, A = 38° 58'$.
9. $a = 4008, c = 8357, B = 61° 21'$.
11. $b = 1.468, A = 26° 5', B = 63° 55'$.
13. $c = 90.47, A = 23° 27', B = 66° 33'$.
15. $a = 0.004293, A = 0.833$ radians, $B = 0.738$ radians.
17. 193.7 feet.   19. 736.2 feet.

Assignment 16, Art. 38, page 62

1. $b = 53.48, c = 54.30, C = 67° 23'$.
3. $a = 1222, c = 1297, C = 75° 33'$.
5. $a = 9.368, b = 0.1810, C = 110° 17'$.
7. $b = 4017, c = 2217, B = 85° 11'$.
9. 842.4 yards.

Assignment 17, Art. 39, page 66

1. $b = 675.8, B = 100° 2', C = 39° 46'$.
3. $\begin{cases} a = 446.2 & A = 34° 0' & C = 80° 45' \\ a' = 213.2 & A' = 15° 30' & C' = 99° 15' \end{cases}$
5. $B = 90° 0', C = 22° 44', c = 481.7$.
7. $\begin{cases} c = 2218 & B = 85° 10' & C = 33° 23' \\ c' = 1621 & B' = 94° 50' & C' = 23° 43' \end{cases}$
9. $b = 2803, A = 14° 29', B = 135° 31'$.   11. 437.2.

Assignment 18, Art. 42, page 70

1. $c = 6.76$.   3. $a = 123.3$.
5. $c = 123.2, A = 55° 36', B = 80° 40'$.
7. $a = 10350, B = 59° 18', C = 53° 22'$.
9. $a = 104.3, B = 13° 51', C = 67° 21'$.
11. 16.51 miles.

Assignment 19, Art. 44, page 73

1. 964.1.   3. 172700.   5. 68.37.   7. 1,429,000.

Assignment 20, Art. 45, pages 75, 76

1. $A = 58° 2', B = 59° 58', C = 62° 0'$, area $= 4327$.
3. $A = 58° 42', B = 78° 18'\ C = 43° 0'$, area $= 119.0$.
5. $A = 41° 24', B = 55° 46', C = 82° 48'$, area $= 9.924$.
7. $52° 48', 127° 12'$.   9. $11° 52'$ north of east (or west).

ANSWERS

SUPPLEMENTARY EXERCISES AND PROBLEMS

Assignment 1, page 77

11. $-150°, -82°, -45°, -250°$.   **13.** $\dfrac{\pi}{3} = 1.0472$ radians.

Assignment 2, pages 77, 78

1. $\sin A = \tfrac{5}{13}$, $\tan A = \tfrac{5}{12}$, $\cot A = \tfrac{12}{5}$, $\sec A = \tfrac{13}{12}$, $\csc A = \tfrac{13}{5}$.

3. $\begin{cases} \sin A = \cos B = \tfrac{2}{3};\ \cos A = \sin B = \tfrac{1}{3}\sqrt{5};\ \tan A = \cot B = \tfrac{2}{5}\sqrt{5}; \\ \cot A = \tan B = \tfrac{1}{2}\sqrt{5};\ \sec A = \csc B = \tfrac{3}{5}\sqrt{5};\ \csc A = \sec B = \tfrac{3}{2}. \end{cases}$

5. $\sin A = \cos B = \tfrac{9}{11}$; $\cos A = \sin B = \tfrac{2}{11}\sqrt{10}$;
$\tan A = \cot B = \tfrac{9}{20}\sqrt{10}$; $\cot A = \tan B = \tfrac{2}{9}\sqrt{10}$;
$\sec A = \csc B = \tfrac{11}{20}\sqrt{10}$; $\csc A = \sec B = \tfrac{11}{9}$.

7. $\sin A = \cos B = \tfrac{3}{5}$; $\cos A = \sin B = \tfrac{4}{5}$; $\tan A = \cot B = \tfrac{3}{4}$;
$\cot A = \tan B = \tfrac{4}{3}$; $\sec A = \csc B = \tfrac{5}{4}$; $\csc A = \sec B = \tfrac{5}{3}$.

9. $\sin A = \dfrac{1}{2}$, $\cos A = \dfrac{\sqrt{3}}{2}$, $\tan A = \dfrac{1}{3}\sqrt{3}$, $\cot A = \sqrt{3}$, $\sec A = \dfrac{2}{3}\sqrt{3}$, $\csc A = 2$.

Assignment 3, page 78

1. Yes; $\sin 45° = \cos 45°$; $\sin 225° = \cos 225°$.
3. $\tan A = \pm \tfrac{1}{3}\sqrt{3}$; $A = 30°$ or $150°$.   **9.** $x = 7°\,30'$.

Assignment 5, page 79

1. $A = 51°\,48'$, $B = 38°\,12'$, $c = 29.3$, area $= 208.2$.
3. $B = 52°\,32'$, $b = 160.5$, $c = 202.2$, area $= 9871$.
5. $A = 51°\,53'$, $b = 27.93$, $c = 45.25$ area $= 497.2$.
7. $AB = 6.560$, $AC = 16.83$, $C = 22°\,0'$, area $= 43.5$.
9. $35°\,16'$.

Assignment 6, page 80

1.

	sin	cos	tan	cot	sec	csc
P_1	$\tfrac{4}{5}$	$-\tfrac{3}{5}$	$-\tfrac{4}{3}$	$-\tfrac{3}{4}$	$-\tfrac{5}{3}$	$\tfrac{5}{4}$
P_2	$-\tfrac{3}{5}$	$\tfrac{4}{5}$	$-\tfrac{3}{4}$	$-\tfrac{4}{3}$	$\tfrac{5}{4}$	$-\tfrac{5}{3}$
P_3	$-\dfrac{1}{3}$	$-\dfrac{2\sqrt{2}}{3}$	$\tfrac{1}{4}\sqrt{2}$	$2\sqrt{2}$	$-\tfrac{3}{4}\sqrt{2}$	-3
P_4	$-\dfrac{9}{\sqrt{130}}$	$-\dfrac{7}{\sqrt{130}}$	$\tfrac{9}{7}$	$\tfrac{7}{9}$	$-\dfrac{\sqrt{130}}{7}$	$-\dfrac{\sqrt{130}}{9}$

3. Two angles, 36° 52′ and 143° 8′. Sine and cosecant are the same for both angles. Cosine, tangent, secant and cotangent differ in sign.

5. $P_1(5, 126° 52')$, $P_2(15, 323° 8')$, $P_3(3, 199° 28')$, $P_4(\sqrt{130}, 232° 8')$.

Assignment 7, page 80

1. $\sin 71°$. **3.** $\tan 83°$. **5.** $-\cos 75°$. **7.** $-\csc 62°$. **9.** $-\sin 62°$.

Assignment 8, pages 80, 81

9. 0°, 60°, 180°, 300°. **11.** 45°.

Assignment 9, page 81

1. $\sin (x+y) = -\dfrac{12+3\sqrt{7}}{20}$, $\cos (x+y) = \dfrac{4\sqrt{7}-9}{20}$,

$\tan (x+y) = -\dfrac{12+3\sqrt{7}}{4\sqrt{7}-9}$.

5. $\cot (x+y) = \dfrac{\cot x \cot y - 1}{\cot x + \cot y}$.

9. $\sin 60° = \dfrac{\sqrt{3}}{2}$.

Assignment 10, pages 81, 82

1. $\sin 150° = \tfrac{1}{2}$, $\cos 150° = -\dfrac{\sqrt{3}}{2}$. **3.** $\cos 2x = -\dfrac{119}{169}$.

7. $\sin \dfrac{x}{2} = \dfrac{5}{\sqrt{26}}$, $\cos \dfrac{x}{2} = \dfrac{1}{\sqrt{26}}$, $\tan \dfrac{x}{2} = 5$.

9. $\sin 135° = \tfrac{1}{2}\sqrt{2}$, $\cos 135° = -\tfrac{1}{2}\sqrt{2}$.

Assignment 11, page 82

3. $\sin 10\theta - \sin 6\theta$. **5.** $\cos x(1 + \sin x)$. **9.** $\sqrt{3}$.

Assignment 12, page 83

1. (a) $10^{0.0792} = 12$; (b) $10^0 = 1$; (c) $10^1 = 10$; (d) $10^{-4} = 0.0001$.

5. (a) $x = 100$; (b) $x = 2$; (c) $x = 7$; (d) $x = 2$; (e) $x = -1$; (f) $x = 1.3010$.

7. $\tfrac{1}{2}[\log 3 + \log 7 - \log 5]$.

9. $2 \log 2 + \log 3 + 2 \log 5 + \log 7 + \log 11$.

Assignment 13, page 83

1. 3.8551. **3.** $9.8566 - 10$.

5. 4.4538, 4.4538, 4.4538, 4.4538, 4.4539.

7. $x = 827.2$. **9.** $z = 1.322$.

ANSWERS

Assignment 14, page 84
1. 0.002081. **3.** 1.259. **5.** 2.405. **7.** $1479; $2188. **9.** 28.11.

Assignment 15, pages 84, 85
1. $A = 5°\ 44'$, $B = 84°\ 16'$, $b = 9.950$.
3. $a = 180.0$, $b = 70.00$, $B = 21°\ 15'$.
5. $8°\ 32'$. **7.** A decreased by $1°\ 6'$. **9.** A increased by $1°\ 5'$.

Assignment 16, page 85
1. $b = 1129$, $c = 1041$, $B = 60°$.
3. $a = 1.169$, $b = 0.7631$, $C = 40°$.
5. $AD = 5.647$, $CB = 8.120$.
7. $a = 11.43$, $b = 21.31$, $B = 167°\ 49'$. **9.** 33010 feet.

Assignment 17, pages 85, 86
1. $A = 21°\ 33'$, $C = 15°\ 27'$, $c = 1259$. **3.** No solution.
5. $b = 14.98$, $B = 90°\ 56'$, $C = 41°\ 51'$.
7. $a = 1017$, $A = 7°\ 34'$, $B = 7°\ 26'$. **9.** $31°\ 40'$.

Assignment 18, page 86
1. 19.70. **2.** $A = 12°\ 22'$, $C = 18°\ 8'$, $b = 6424$.
5. $a = 3021$, $B = 3°\ 18'$, $C = 166°\ 42'$.
7. $A = 7°\ 30'$, $B = 0°\ 45'$, $c = 3738$. **9.** 7.280 inches; 12.72 inches.

Assignment 19, page 87
1. 45220. **3.** 8676. **5.** 30070.
9. $\dfrac{b^2 - a^2}{2} \tan A$, where $b =$ the longer, a the shorter of the parallel sides.

Assignment 20, page 87
1. $A = 50°\ 28'$, $B = 59°\ 0'$, $C = 70°\ 32'$; area $= 42.43$.
3. $A = 49°\ 2'$, $B = 76°\ 6'$, $C = 54°\ 52'$; area $= 803200$.
5. $A = 36°\ 52'$, $B = 53°\ 8'$, $C = 90°$; area $= 60000$.
7. $r = 1$. **9.** 261.9 chains.

MISCELLANEOUS EXERCISES, PAGES 89-93

1. 242°, 39°, 206°, 45°, 150°, 307°.
3. 36°, 25$\frac{5}{7}$°, 40°, 10°, 110°, 12°, 63°, 6°.   **5.** 18.5 miles per second.
7. $\begin{cases} A \text{ in 3rd quadrant, } \sin A = -\frac{8}{17}, \cos A = -\frac{15}{17}, \tan A = \frac{8}{15}, \\ \quad \cot A = \frac{15}{8}, \sec A = -\frac{17}{15}. \\ A \text{ in 4th quadrant, } \sin A = -\frac{8}{17}, \cos A = \frac{15}{17}, \tan A = -\frac{8}{15}, \\ \quad \cot A = -\frac{15}{8}, \sec A = \frac{17}{15}. \end{cases}$

11. $\cos A = \sqrt{1 - \sin^2 A}$, $\tan A = \dfrac{\sin A}{\sqrt{1 - \sin^2 A}}$,

$\cot A = \dfrac{\sqrt{1 - \sin^2 A}}{\sin A}$, $\sec A = \dfrac{1}{\sqrt{1 - \sin^2 A}}$, $\csc A = \dfrac{1}{\sin A}$.

23. 5.16 square inches.   **25.** $CA = 1098$ feet.
27. $(-5, 0), (-4, 0), (-2, -2\sqrt{3}), (0,0), (0, -5)$.
29. $\theta = 36°\, 52'$; $\sin \theta = \frac{3}{5}$, $\cos \theta = \frac{4}{5}$, $\tan \theta = \frac{3}{4}$, $\cot \theta = \frac{4}{3}$, $\sec \theta = \frac{5}{4}$, $\csc \theta = \frac{5}{3}$.
31. $3°\, 11'$.   **33.** $-\sin 27°$.   **35.** $-\csc 28°$.   **37.** $\sec 32°$.
39. $-\cos 10°$.   **41.** $-\cot 19°$.   **51.** $x = 60°$.   **53.** $x = 60°$ or $300°$.
55. $x = 30°, 90°$ or $150°$.   **57.** $\log 3$.   **59.** -6.   **61.** 3.066.   **63.** 0.7943.
65. 39.95 miles.   **71.** 38.14 feet.   **73.** 8.932, 20.18 inches.   **81.** 2011.

Tables

I. Four-Place Logarithms

II. Four-Place Values of Functions and Radians

III. Logarithms of Trigonometric Functions

IV. Constants and their Logarithms

I. FOUR-PLACE LOGARITHMS

n	0	1	2	3	4	5	6	7	8	9
10	0000	0043	0086	0128	0170	0212	0253	0294	0334	0374
11	0414	0453	0492	0531	0569	0607	0645	0682	0719	0755
12	0792	0828	0864	0899	0934	0969	1004	1038	1072	1106
13	1139	1173	1206	1239	1271	1303	1335	1367	1399	1430
14	1461	1492	1523	1553	1584	1614	1644	1673	1703	1732
15	1761	1790	1818	1847	1875	1903	1931	1959	1987	2014
16	2041	2068	2095	2122	2148	2175	2201	2227	2253	2279
17	2304	2330	2355	2380	2405	2430	2455	2480	2504	2529
18	2553	2577	2601	2625	2648	2672	2695	2718	2742	2765
19	2788	2810	2833	2856	2878	2900	2923	2945	2967	2989
20	3010	3032	3054	3075	3096	3118	3139	3160	3181	3201
21	3222	3243	3263	3284	3304	3324	3345	3365	3385	3404
22	3424	3444	3464	3483	3502	3522	3541	3560	3579	3598
23	3617	3636	3655	3674	3692	3711	3729	3747	3766	3784
24	3802	3820	3838	3856	3874	3892	3909	3927	3945	3962
25	3979	3997	4014	4031	4048	4065	4082	4099	4116	4133
26	4150	4166	4183	4200	4216	4232	4249	4265	4281	4298
27	4314	4330	4346	4362	4378	4393	4409	4425	4440	4456
28	4472	4487	4502	4518	4533	4548	4564	4579	4594	4609
29	4624	4639	4654	4669	4683	4698	4713	4728	4742	4757
30	4771	4786	4800	4814	4829	4843	4857	4871	4886	4900
31	4914	4928	4942	4955	4969	4983	4997	5011	5024	5038
32	5051	5065	5079	5092	5105	5119	5132	5145	5159	5172
33	5185	5198	5211	5224	5237	5250	5263	5276	5289	5302
34	5315	5328	5340	5353	5366	5378	5391	5403	5416	5428
35	5441	5453	5465	5478	5490	5502	5514	5527	5539	5551
36	5563	5575	5587	5599	5611	5623	5635	5647	5658	5670
37	5682	5694	5705	5717	5729	5740	5752	5763	5775	5786
38	5798	5809	5821	5832	5843	5855	5866	5877	5888	5899
39	5911	5922	5933	5944	5955	5966	5977	5988	5999	6010
40	6021	6031	6042	6053	6064	6075	6085	6096	6107	6117
41	6128	6138	6149	6160	6170	6180	6191	6201	6212	6222
42	6232	6243	6253	6263	6274	6284	6294	6304	6314	6325
43	6335	6345	6355	6365	6375	6385	6395	6405	6415	6425
44	6435	6444	6454	6464	6474	6484	6493	6503	6513	6522
45	6532	6542	6551	6561	6571	6580	6590	6599	6609	6618
46	6628	6637	6646	6656	6665	6675	6684	6693	6702	6712
47	6721	6730	6739	6749	6758	6767	6776	6785	6794	6803
48	6812	6821	6830	6839	6848	6857	6866	6875	6884	6893
49	6902	6911	6920	6928	6937	6946	6955	6964	6972	6981
50	6990	6998	7007	7016	7024	7033	7042	7050	7059	7067
51	7076	7084	7093	7101	7110	7118	7126	7135	7143	7152
52	7160	7168	7177	7185	7193	7202	7210	7218	7226	7235
53	7243	7251	7259	7267	7275	7284	7292	7300	7308	7316
54	7324	7332	7340	7348	7356	7364	7372	7380	7388	7396

I. FOUR-PLACE LOGARITHMS (Continued)

n	0	1	2	3	4	5	6	7	8	9
55	7404	7412	7419	7427	7435	7443	7451	7459	7466	7474
56	7482	7490	7497	7505	7513	7520	7528	7536	7543	7551
57	7559	7566	7574	7582	7589	7597	7604	7612	7619	7627
58	7634	7642	7649	7657	7664	7672	7679	7686	7694	7701
59	7709	7716	7723	7731	7738	7745	7752	7760	7767	7774
60	7782	7789	7796	7803	7810	7818	7825	7832	7839	7846
61	7853	7860	7868	7875	7882	7889	7896	7903	7910	7917
62	7924	7931	7938	7945	7952	7959	7966	7973	7980	7987
63	7993	8000	8007	8014	8021	8028	8035	8041	8048	8055
64	8062	8069	8075	8082	8089	8096	8102	8109	8116	8122
65	8129	8136	8142	8149	8156	8162	8169	8176	8182	8189
66	8195	8202	8209	8215	8222	8228	8235	8241	8248	8254
67	8261	8267	8274	8280	8287	8293	8299	8306	8312	8319
68	8325	8331	8338	8344	8351	8357	8363	8370	8376	8382
69	8388	8395	8401	8407	8414	8420	8426	8432	8439	8445
70	8451	8457	8463	8470	8476	8482	8488	8494	8500	8506
71	8513	8519	8525	8531	8537	8543	8549	8555	8561	8567
72	8573	8579	8585	8591	8597	8603	8609	8615	8621	8627
73	8633	8639	8645	8651	8657	8663	8669	8675	8681	8686
74	8692	8698	8704	8710	8716	8722	8727	8733	8739	8745
75	8751	8756	8762	8768	8774	8779	8785	8791	8797	8802
76	8808	8814	8820	8825	8831	8837	8842	8848	8854	8859
77	8865	8871	8876	8882	8887	8893	8899	8904	8910	8915
78	8921	8927	8932	8938	8943	8949	8954	8960	8965	8971
79	8976	8982	8987	8993	8998	9004	9009	9015	9020	9025
80	9031	9036	9042	9047	9053	9058	9063	9069	9074	9079
81	9085	9090	9096	9101	9106	9112	9117	9122	9128	9133
82	9138	9143	9149	9154	9159	9165	9170	9175	9180	9186
83	9191	9196	9201	9206	9212	9217	9222	9227	9232	9238
84	9243	9248	9253	9258	9263	9269	9274	9279	9284	9289
85	9294	9299	9304	9309	9315	9320	9325	9330	9335	9340
86	9345	9350	9355	9360	9365	9370	9375	9380	9385	9390
87	9395	9400	9405	9410	9415	9420	9425	9430	9435	9440
88	9445	9450	9455	9460	9465	9469	9474	9479	9484	9489
89	9494	9499	9504	9509	9513	9518	9523	9528	9533	9538
90	9542	9547	9552	9557	9562	9566	9571	9576	9581	9586
91	9590	9595	9600	9605	9609	9614	9619	9624	9628	9633
92	9638	9643	9647	9652	9657	9661	9666	9671	9675	9680
93	9685	9689	9694	9699	9703	9708	9713	9717	9722	9727
94	9731	9736	9741	9745	9750	9754	9759	9763	9768	9773
95	9777	9782	9786	9791	9795	9800	9805	9809	9814	9818
96	9823	9827	9832	9836	9841	9845	9850	9854	9859	9863
97	9868	9872	9877	9881	9886	9890	9894	9899	9903	9908
98	9912	9917	9921	9926	9930	9934	9939	9943	9948	9952
99	9956	9961	9965	9969	9974	9978	9983	9987	9991	9996

II. FOUR-PLACE VALUES OF FUNCTIONS AND RADIANS

Degrees	Radians	Sin	Csc	Tan	Cot	Sec	Cos		
0° 0′	.0000	.0000	—	.0000	—	1.000	1.0000	1.5708	90° 0′
10′	029	029	343.8	029	343.8	000	000	679	50′
20′	058	058	171.9	058	171.9	000	000	650	40′
30′	.0087	.0087	114.6	.0087	114.6	1.000	1.0000	1.5621	30′
40′	116	116	85.95	116	85.94	000	.9999	592	20′
50′	145	145	68.76	145	68.75	000	999	563	10′
1° 0′	.0175	.0175	57.30	.0175	57.29	1.000	.9998	1.5533	89° 0′
10′	204	204	49.11	204	49.10	000	998	504	50′
20′	233	233	42.98	233	42.96	000	997	475	40′
30′	.0262	.0262	38.20	.0262	38.19	1.000	.9997	1.5446	30′
40′	291	291	34.38	291	34.37	000	996	417	20′
50′	320	320	31.26	320	31.24	001	995	388	10′
2° 0′	.0349	.0349	28.65	.0349	28.64	1.001	.9994	1.5359	88° 0′
10′	378	378	26.45	378	26.43	001	993	330	50′
20′	407	407	24.56	407	24.54	001	992	301	40′
30′	.0436	.0436	22.93	.0437	22.90	1.001	.9990	1.5272	30′
40′	465	465	21.49	466	21.47	001	989	243	20′
50′	495	494	20.23	495	20.21	001	988	213	10′
3° 0′	.0524	.0523	19.11	.0524	19.08	1.001	.9986	1.5184	87° 0′
10′	553	552	18.10	553	18.07	002	985	155	50′
20′	582	581	17.20	582	17.17	002	983	126	40′
30′	.0611	.0610	16.38	.0612	16.35	1.002	.9981	1.5097	30′
40′	640	640	15.64	641	15.60	002	980	068	20′
50′	669	669	14.96	670	14.92	002	978	039	10′
4° 0′	.0698	.0698	14.34	.0699	14.30	1.002	.9976	1.5010	86° 0′
10′	727	727	13.76	729	13.73	003	974	981	50′
20′	756	756	13.23	758	13.20	003	971	952	40′
30′	.0785	.0785	12.75	.0787	12.71	1.003	.9969	1.4923	30′
40′	814	814	12.29	816	12.25	003	967	893	20′
50′	844	843	11.87	846	11.83	004	964	864	10′
5° 0′	.0873	.0872	11.47	.0875	11.43	1.004	.9962	1.4835	85° 0′
10′	902	901	11.10	904	11.06	004	959	806	50′
20′	931	929	10.76	934	10.71	004	957	777	40′
30′	.0960	.0958	10.43	.0963	10.39	1.005	.9954	1.4748	30′
40′	989	987	10.13	992	10.08	005	951	719	20′
50′	.1018	.1016	9.839	.1022	9.788	005	948	690	10′
6° 0′	.1047	.1045	9.567	.1051	9.514	1.006	.9945	1.4661	84° 0′
10′	076	074	9.309	080	9.255	006	942	632	50′
20′	105	103	9.065	110	9.010	006	939	603	40′
30′	.1134	.1132	8.834	.1139	8.777	1.006	.9936	1.4573	30′
40′	164	161	8.614	169	8.556	007	932	544	20′
50′	193	190	8.405	198	8.345	007	929	515	10′
7° 0′	.1222	.1219	8.206	.1228	8.144	1.008	.9925	1.4486	83° 0′
10′	251	248	8.016	257	7.953	008	922	457	50′
20′	280	276	7.834	287	7.770	008	918	428	40′
30′	.1309	.1305	7.661	.1317	7.596	1.009	.9914	1.4399	30′
40′	338	334	7.496	346	7.429	009	911	370	20′
50′	367	363	7.337	376	7.269	009	907	341	10′
8° 0′	.1396	.1392	7.185	.1405	7.115	1.010	.9903	1.4312	82° 0′
10′	425	421	7.040	435	6.968	010	899	283	50′
20′	454	449	6.900	465	6.827	011	894	254	40′
30′	.1484	.1478	6.765	.1495	6.691	1.011	.9890	1.4224	30′
40′	513	507	6.636	524	6.561	012	886	195	20′
50′	542	536	6.512	554	6.435	012	881	166	10′
9° 0′	.1571	.1564	6.392	.1584	6.314	1.012	.9877	1.4137	81° 0′
		Cos	Sec	Cot	Tan	Csc	Sin	Radians	Degrees

II. FOUR-PLACE VALUES OF FUNCTIONS AND RADIANS (Cont.)

Degrees	Radians	Sin	Csc	Tan	Cot	Sec	Cos		
9° 0'	.1571	.1564	6.392	.1584	6.314	1.012	.9877	1.4137	81° 0'
10'	600	593	277	614	197	013	872	108	50'
20'	629	622	166	644	084	013	868	079	40'
30'	.1658	.1650	6.059	.1673	5.976	1.014	.9863	1.4050	30'
40'	687	679	5.955	703	871	014	858	1.4021	20'
50'	716	708	855	733	769	015	853	992	10'
10° 0'	.1745	.1736	5.759	.1763	5.671	1.015	.9848	1.3963	80° 0'
10'	774	765	665	793	576	016	843	934	50'
20'	804	794	575	823	485	016	838	904	40'
30'	.1833	.1822	5.487	.1853	5.396	1.017	.9833	1.3875	30'
40'	862	851	403	883	309	018	827	846	20'
50'	891	880	320	914	226	018	822	817	10'
11° 0'	.1920	.1908	5.241	.1944	5.145	1.019	.9816	1.3788	79° 0'
10'	949	937	164	974	066	019	811	759	50'
20'	978	965	089	.2004	4.989	020	805	730	40'
30'	.2007	.1994	5.016	.2035	4.915	1.020	.9799	1.3701	30'
40'	036	.2022	4.945	065	843	021	793	672	20'
50'	065	051	876	095	773	022	787	643	10'
12° 0'	.2094	.2079	4.810	.2126	4.705	1.022	.9781	1.3614	78° 0'
10'	123	108	745	156	638	023	775	584	50'
20'	153	136	682	186	574	024	769	555	40'
30'	.2182	.2164	4.620	.2217	4.511	1.024	.9763	1.3526	30'
40'	211	193	560	247	449	025	757	497	20'
50'	240	221	502	278	390	026	750	468	10'
13° 0'	.2269	.2250	4.445	.2309	4.331	1.026	.9744	1.3439	77° 0'
10'	298	278	390	339	275	027	737	410	50'
20'	327	306	336	370	219	028	730	381	40'
30'	.2356	.2334	4.284	.2401	4.165	1.028	.9724	1.3352	30'
40'	385	363	232	432	113	029	717	323	20'
50'	414	391	182	462	061	030	710	294	10'
14° 0'	.2443	.2419	4.134	.2493	4.011	1.031	.9703	1.3265	76° 0'
10'	473	447	086	524	3.962	031	696	235	50'
20'	502	476	039	555	914	032	689	206	40'
30'	.2531	.2504	3.994	.2586	3.867	1.033	.9681	1.3177	30'
40'	560	532	950	617	821	034	674	148	20'
50'	589	560	906	648	776	034	667	119	10'
15° 0'	.2618	.2588	3.864	.2679	3.732	1.035	.9659	1.3090	75° 0'
10'	647	616	822	711	689	036	652	061	50'
20'	676	644	782	742	647	037	644	032	40'
30'	.2705	.2672	3.742	.2773	3.606	1.038	.9636	1.3003	30'
40'	734	700	703	805	566	039	628	974	20'
50'	763	728	665	836	526	039	621	945	10'
16° 0'	.2793	.2756	3.628	.2867	3.487	1.040	.9613	1.2915	74° 0'
10'	822	784	592	899	450	041	605	886	50'
20'	851	812	556	931	412	042	596	857	40'
30'	.2880	.2840	3.521	.2962	3.376	1.043	.9588	1.2828	30'
40'	909	868	487	994	340	044	580	799	20'
50'	938	896	453	.3026	305	045	572	770	10'
17° 0'	.2967	.2924	3.420	.3057	3.271	1.046	.9563	1.2741	73° 0'
10'	996	952	388	089	237	047	555	712	50'
20'	.3025	979	357	121	204	048	546	683	40'
30'	.3054	.3007	3.326	.3153	3.172	1.048	.9537	1.2654	30'
40'	083	035	295	185	140	049	528	625	20'
50'	113	062	265	217	108	050	520	595	10'
18° 0'	.3142	.3090	3.236	.3249	3.078	1.051	.9511	1.2566	72° 0'
		Cos	Sec	Cot	Tan	Csc	Sin	Radians	Degrees

II. FOUR-PLACE VALUES OF FUNCTIONS AND RADIANS (Cont.)

Degrees	Radians	Sin	Csc	Tan	Cot	Sec	Cos	Radians	
18° 0'	.3142	.3090	3.236	.3249	3.078	1.051	.9511	1.2566	72° 0'
10'	171	118	207	281	047	052	502	537	50'
20'	200	145	179	314	018	053	492	508	40'
30'	.3229	.3173	3.152	.3346	2.989	1.054	.9483	1.2479	30'
40'	258	201	124	378	960	056	474	450	20'
50'	287	228	098	411	932	057	465	421	10'
19° 0'	.3316	.3256	3.072	.3443	2.904	1.058	.9455	1.2392	71° 0'
10'	345	283	046	476	877	059	446	363	50'
20'	374	311	021	508	850	060	436	334	40'
30'	.3403	.3338	2.996	.3541	2.824	1.061	.9426	1.2305	30'
40'	432	365	971	574	798	062	417	275	20'
50'	462	393	947	607	773	063	407	246	10'
20° 0'	.3491	.3420	2.924	.3640	2.747	1.064	.9397	1.2217	70° 0'
10'	520	448	901	673	723	065	387	188	50'
20'	549	475	878	706	699	066	377	159	40'
30'	.3578	.3502	2.855	.3739	2.675	1.068	.9367	1.2130	30'
40'	607	529	833	772	651	069	356	101	20'
50'	636	557	812	805	628	070	346	072	10'
21° 0'	.3665	.3584	2.790	.3839	2.605	1.071	.9336	1.2043	69° 0'
10'	694	611	769	872	583	072	325	1.2014	50'
20'	723	638	749	906	560	074	315	985	40'
30'	.3752	.3665	2.729	.3939	2.539	1.075	.9304	1.1956	30'
40'	782	692	709	973	517	076	293	926	20'
50'	811	719	689	.4006	496	077	283	897	10'
22° 0'	.3840	.3746	2.669	.4040	2.475	1.079	.9272	1.1868	68° 0'
10'	869	773	650	074	455	080	261	839	50'
20'	898	800	632	108	434	081	250	810	40'
30'	.3927	.3827	2.613	.4142	2.414	1.082	.9239	1.1781	30'
40'	956	854	595	176	394	084	228	752	20'
50'	985	881	577	210	375	085	216	723	10'
23° 0'	.4014	.3907	2.559	.4245	2.356	1.086	.9205	1.1694	67° 0'
10'	043	934	542	279	337	088	194	665	50'
20'	072	961	525	314	318	089	182	636	40'
30'	.4102	.3987	2.508	.4348	2.300	1.090	.9171	1.1606	30'
40'	131	.4014	491	383	282	092	159	577	20'
50'	160	041	475	417	264	093	147	548	10'
24° 0'	.4189	.4067	2.459	.4452	2.246	1.095	.9135	1.1519	66° 0'
10'	218	094	443	487	229	096	124	490	50'
20'	247	120	427	522	211	097	112	461	40'
30'	.4276	.4147	2.411	.4557	2.194	1.099	.9100	1.1432	30'
40'	305	173	396	592	177	100	088	403	20'
50'	334	200	381	628	161	102	075	374	10'
25° 0'	.4363	.4226	2.366	.4663	2.145	1.103	.9063	1.1345	65° 0'
10'	392	253	352	699	128	105	051	316	50'
20'	422	279	337	734	112	106	038	286	40'
30'	.4451	.4305	2.323	.4770	2.097	1.108	.9026	1.1257	30'
40'	480	331	309	806	081	109	013	228	20'
50'	509	358	295	841	066	111	001	199	10'
26° 0'	.4538	.4384	2.281	.4877	2.050	1.113	.8988	1.1170	64° 0'
10'	567	410	268	913	035	114	975	141	50'
20'	596	436	254	950	020	116	962	112	40'
30'	.4625	.4462	2.241	.4986	2.006	1.117	.8949	1.1083	30'
40'	654	488	228	.5022	1.991	119	936	054	20'
50'	683	514	215	059	977	121	923	1.1025	10'
27° 0'	.4712	.4540	2.203	.5095	1.963	1.122	.8910	1.0996	63° 0'
		Cos	Sec	Cot	Tan	Csc	Sin	Radians	Degrees

— 110 —

II. FOUR-PLACE VALUES OF FUNCTIONS AND RADIANS (Cont.)

Degrees	Radians	Sin	Csc	Tan	Cot	Sec	Cos		
27° 0'	.4712	.4540	2.203	.5095	1.963	1.122	.8910	1.0996	63° 0'
10'	741	566	190	132	949	124	897	966	50'
20'	771	592	178	169	935	126	884	937	40'
30'	.4800	.4617	2.166	.5206	1.921	1.127	.8870	1.0908	30'
40'	829	643	154	243	907	129	857	879	20'
50'	858	669	142	280	894	131	843	850	10'
28° 0'	.4887	.4695	2.130	.5317	1.881	1.133	.8829	1.0821	62° 0'
10'	916	720	118	354	868	134	816	792	50'
20'	945	746	107	392	855	136	802	763	40'
30'	.4974	.4772	2.096	.5430	1.842	1.138	.8788	1.0734	30'
40'	.5003	797	085	467	829	140	774	705	20'
50'	032	823	074	505	816	142	760	676	10'
29° 0'	.5061	.4848	2.063	.5543	1.804	1.143	.8746	1.0647	61° 0'
10'	091	874	052	581	792	145	732	617	50'
20'	120	899	041	619	780	147	718	588	40'
30'	.5149	.4924	2.031	.5658	1.767	1.149	.8704	1.0559	30'
40'	178	950	020	696	756	151	689	530	20'
50'	207	975	010	735	744	153	675	501	10'
30° 0'	.5236	.5000	2.000	.5774	1.732	1.155	.8660	1.0472	60° 0'
10'	265	025	1.990	812	720	157	646	443	50'
20'	294	050	980	851	709	159	631	414	40'
30'	.5323	.5075	1.970	.5890	1.698	1.161	.8616	1.0385	30'
40'	352	100	961	930	686	163	601	356	20'
50'	381	125	951	969	675	165	587	327	10'
31° 0'	.5411	.5150	1.942	.6009	1.664	1.167	.8572	1.0297	59° 0'
10'	440	175	932	048	653	169	557	268	50'
20'	469	200	923	088	643	171	542	239	40'
30'	.5498	.5225	1.914	.6128	1.632	1.173	.8526	1.0210	30'
40'	527	250	905	168	621	175	511	181	20'
50'	556	275	896	208	611	177	496	152	10'
32° 0'	.5585	.5299	1.887	.6249	1.600	1.179	.8480	1.0123	58° 0'
10'	614	324	878	289	590	181	465	094	50'
20'	643	348	870	330	580	184	450	065	40'
30'	.5672	.5373	1.861	.6371	1.570	1.186	.8434	1.0036	30'
40'	701	398	853	412	560	188	418	1.0007	20'
50'	730	422	844	453	550	190	403	977	10'
33° 0'	.5760	.5446	1.836	.6494	1.540	1.192	.8387	.9948	57° 0'
10'	789	471	828	536	530	195	371	919	50'
20'	818	495	820	577	520	197	355	890	40'
30'	.5847	.5519	1.812	.6619	1.511	1.199	.8339	.9861	30'
40'	876	544	804	661	501	202	323	832	20'
50'	905	568	796	703	1.492	204	307	803	10'
34° 0'	.5934	.5592	1.788	.6745	1.483	1.206	.8290	.9774	56° 0'
10'	963	616	781	787	473	209	274	745	50'
20'	992	640	773	830	464	211	258	716	40'
30'	.6021	.5664	1.766	.6873	1.455	1.213	.8241	.9687	30'
40'	050	688	758	916	446	216	225	657	20'
50'	080	712	751	959	437	218	208	628	10'
35° 0'	.6109	.5736	1.743	.7002	1.428	1.221	.8192	.9599	55° 0'
10'	138	760	736	046	419	223	175	570	50'
20'	167	783	729	089	411	226	158	541	40'
30'	.6196	.5807	1.722	.7133	1.402	1.228	.8141	.9512	30'
40'	225	831	715	177	393	231	124	483	20'
50'	254	854	708	221	385	233	107	454	10'
36° 0'	.6283	.5878	1.701	.7265	1.376	1.236	.8090	.9425	54° 0'
		Cos	Sec	Cot	Tan	Csc	Sin	Radians	Degrees

II. FOUR-PLACE VALUES OF FUNCTIONS AND RADIANS (*Cont.*)

Degrees	Radians	Sin	Csc	Tan	Cot	Sec	Cos		
36° 0'	.6283	.5878	1.701	.7265	1.376	1.236	.8090	.9425	**54° 0'**
10'	312	901	695	310	368	239	073	396	50'
20'	341	925	688	355	360	241	056	367	40'
30'	.6370	.5948	1.681	.7400	1.351	1.244	.8039	.9338	30'
40'	400	972	675	445	343	247	021	308	20'
50'	429	995	668	490	335	249	004	279	10'
37° 0'	.6458	.6018	1.662	.7536	1.327	1.252	.7986	.9250	**53° 0'**
10'	487	041	655	581	319	255	969	221	50'
20'	516	065	649	627	311	258	951	192	40'
30'	.6545	.6088	1.643	.7673	1.303	1.260	.7934	.9163	30'
40'	574	111	636	720	295	263	916	134	20'
50'	603	134	630	766	288	266	898	105	10'
38° 0'	.6632	.6157	1.624	.7813	1.280	1.269	.7880	.9076	**52° 0'**
10'	661	180	618	860	272	272	862	047	50'
20'	690	202	612	907	265	275	844	.9018	40'
30'	.6720	.6225	1.606	.7954	1.257	1.278	.7826	.8988	30'
40'	749	248	601	.8002	250	281	808	959	20'
50'	778	271	595	050	242	284	790	930	10'
39° 0'	.6807	.6293	1.589	.8098	1.235	1.287	.7771	.8901	**51° 0'**
10'	836	316	583	146	228	290	753	872	50'
20'	865	338	578	195	220	293	735	843	40'
30'	.6894	.6361	1.572	.8243	1.213	1.296	.7716	.8814	30'
40'	923	383	567	292	206	299	698	785	20'
50'	952	406	561	342	199	302	679	756	10'
40° 0'	.6981	.6428	1.556	.8391	1.192	1.305	.7660	.8727	**50° 0'**
10'	.7010	450	550	441	185	309	642	698	50'
20'	039	472	545	491	178	312	623	668	40'
30'	.7069	.6494	1.540	.8541	1.171	1.315	.7604	.8639	30'
40'	098	517	535	591	164	318	585	610	20'
50'	127	539	529	642	157	322	566	581	10'
41° 0'	.7156	.6561	1.524	.8693	1.150	1.325	.7547	.8552	**49° 0'**
10'	185	583	519	744	144	328	528	523	50'
20'	214	604	514	796	137	332	509	494	40'
30'	.7243	.6626	1.509	.8847	1.130	1.335	.7490	.8465	30'
40'	272	648	504	899	124	339	470	436	20'
50'	301	670	499	952	117	342	451	407	10'
42° 0'	.7330	.6691	1.494	.9004	1.111	1.346	.7431	.8378	**48° 0'**
10'	359	713	490	057	104	349	412	348	50'
20'	389	734	485	110	098	353	392	319	40'
30'	.7418	.6756	1.480	.9163	1.091	1.356	.7373	.8290	30
40'	447	777	476	217	085	360	353	261	20'
50'	476	799	471	271	079	364	333	232	10'
43° 0'	.7505	.6820	1.466	.9325	1.072	1.367	.7314	.8203	**47° 0'**
10'	534	841	462	380	066	371	294	174	50'
20'	563	862	457	435	060	375	274	145	40'
30'	.7592	.6884	1.453	.9490	1.054	1.379	.7254	.8116	30'
40'	621	905	448	545	048	382	234	087	20'
50'	650	926	444	601	042	386	214	058	10'
44° 0'	.7679	.6947	1.440	.9657	1.036	1.390	.7193	.8029	**46° 0'**
10'	709	967	435	713	030	394	173	999	50'
20'	738	988	431	770	024	398	153	970	40'
30'	.7767	.7009	1.427	.9827	1.018	1.402	.7133	.7941	30'
40'	796	030	423	884	012	406	112	912	20'
50'	825	050	418	942	006	410	092	883	10'
45° 0'	.7854	.7071	1.414	1.000	1.000	1.414	.7071	.7854	**45° 0'**
		Cos	Sec	Cot	Tan	Csc	Sin	Radians	Degrees

III. LOGARITHMS OF TRIGONOMETRIC FUNCTIONS*

Angle	L Sin	d 1'	L Tan	cd 1'	L Cot	d 1'	L Cos	
0° 0'						.0	10.0000	90° 0'
10'	7.4637		7.4637		12.5363	.0	.0000	50'
20'	.7648	301.1	.7648	301.1	.2352	.0	.0000	40'
		176.0		176.1		.0		
30'	.9408	125.0	.9409	124.9	.0591	.0	.0000	30'
40'	8.0658	96.9	8.0658	96.9	11.9342	.0	.0000	20'
50'	.1627	79.2	.1627	79.2	.8373	.0	.0000	10'
1° 0'	8.2419	66.9	8.2419	67.0	11.7581	.1	9.9999	89° 0'
10'	.3088	58.0	.3089	58.0	.6911	.0	.9999	50'
20'	.3668	51.1	.3669	51.2	.6331	.0	.9999	40'
30'	.4179	45.8	.4181	45.7	.5819	.0	.9999	30'
40'	.4637	41.3	.4638	41.5	.5362	.1	.9998	20'
50'	.5050	37.8	.5053	37.8	.4947	.0	.9998	10'
2° 0'	8.5428	34.8	8.5431	34.8	11.4569	.1	9.9997	88° 0'
10'	.5776	32.1	.5779	32.2	.4221	.0	.9997	50'
20'	.6097	30.0	.6101	30.0	.3899	.1	.9996	40'
30'	.6397	28.0	.6401	28.1	.3599	.0	.9996	30'
40'	.6677	26.3	.6682	26.3	.3318	.1	.9995	20'
50'	.6940	24.8	.6945	24.9	.3055	.0	.9995	10'
3° 0'	8.7188	23.5	8.7194	23.5	11.2806	.1	9.9994	87° 0'
10'	.7423	22.2	.7429	22.3	.2571	.1	.9993	50'
20'	.7645	21.2	.7652	21.3	.2348	.0	.9993	40'
30'	.7857	20.2	.7865	20.2	.2135	.1	.9992	30'
40'	.8059	19.2	.8067	19.4	.1933	.1	.9991	20'
50'	.8251	18.5	.8261	18.5	.1739	.1	.9990	10'
4° 0'	8.8436	17.7	8.8446	17.8	11.1554	.1	9.9989	86° 0'
10'	.8613	17.0	.8624	17.1	.1376	.0	.9989	50'
20'	.8783	16.3	.8795	16.5	.1205	.1	.9988	40'
30'	.8946	15.8	.8960	15.8	.1040	.1	.9987	30'
40'	.9104	15.2	.9118	15.4	.0882	.1	.9986	20'
50'	.9256	14.7	.9272	14.8	.0728	.1	.9985	10'
5° 0'	8.9403	14.2	8.9420	14.3	11.0580	.2	9.9983	85° 0'
10'	.9545	13.7	.9563	13.8	.0437	.1	.9982	50'
20'	.9682	13.4	.9701	13.5	.0299	.1	.9981	40'
30'	.9816	12.9	.9836	13.0	.0164	.1	.9980	30'
40'	.9945	12.5	.9966	12.7	.0034	.1	.9979	20'
50'	9.0070	12.2	9.0093	12.3	10.9907	.2	.9977	10'
6° 0'	9.0192	11.9	9.0216	12.0	10.9784	.1	9.9976	84° 0'
10'	.0311	11.5	.0336	11.7	.9664	.1	.9975	50'
20'	.0426	11.3	.0453	11.4	.9547	.2	.9973	40'
30'	.0539	10.9	.0567	11.1	.9433	.1	.9972	30'
40'	.0648	10.7	.0678	10.8	.9322	.1	.9971	20'
50'	.0755	10.4	.0786	10.5	.9214	.2	.9969	10'
7° 0'	9.0859	10.2	9.0891	10.4	10.9109	.1	9.9968	83° 0'
10'	.0961	9.9	.0995	10.1	.9005	.2	.9966	50'
20'	.1060	9.7	.1096	9.8	.8904	.2	.9964	40'
30'	.1157	9.5	.1194	9.7	.8806	.1	.9963	30'
40'	.1252	9.3	.1291	9.4	.8709	.2	.9961	20'
50'	.1345	9.1	.1385	9.3	.8615	.2	.9959	10'
8° 0'	9.1436	8.9	9.1478	9.1	10.8522	.1	9.9958	82° 0'
10'	.1525	8.7	.1569	8.9	.8431	.2	.9956	50'
20'	.1612	8.5	.1658	8.7	.8342	.2	.9954	40'
30'	.1697	8.4	.1745	8.6	.8255	.2	.9952	30'
40'	.1781	8.2	.1831	8.4	.8169	.2	.9950	20'
50'	.1863	8.0	.1915	8.2	.8085	.2	.9948	10'
9° 0'	9.1943		9.1997		10.8003		9.9946	81° 0'
	L Cos	d 1'	L Cot	cd 1'	L Tan	d 1'	L Sin	Angle

*For simplicity, − 10 has been omitted after each entry.

III. LOGARITHMS OF TRIGONOMETRIC FUNCTIONS (Cont.)

Angle	L Sin	d 1'	L Tan	cd 1'	L Cot	d 1'	L Cos	
9° 0'	9.1943		9.1997		10.8003		9.9946	81° 0'
10'	.2022	7.9	.2078	8.1	.7922	.2	.9944	50'
20'	.2100	7.8	.2158	8.0	.7842	.2	.9942	40'
30'	.2176	7.6	.2236	7.8	.7764	.2	.9940	30'
40'	.2251	7.5	.2313	7.7	.7687	.2	.9938	20'
50'	.2324	7.3	.2389	7.6	.7611	.2	.9936	10'
10° 0'	9.2397	7.3	9.2463	7.4	10.7537	.2	9.9934	80° 0'
10'	.2468	7.1	.2536	7.3	.7464	.3	.9931	50'
20'	.2538	7.0	.2609	7.3	.7391	.2	.9929	40'
30'	.2606	6.8	.2680	7.1	.7320	.2	.9927	30'
40'	.2674	6.8	.2750	7.0	.7250	.3	.9924	20'
50'	.2740	6.6	.2819	6.9	.7181	.2	.9922	10'
11° 0'	9.2806	6.6	9.2887	6.8	10.7113	.3	9.9919	79° 0'
10'	.2870	6.4	.2953	6.6	.7047	.2	.9917	50'
20'	.2934	6.4	.3020	6.7	.6980	.3	.9914	40'
30'	.2997	6.3	.3085	6.5	.6915	.2	.9912	30'
40'	.3058	6.1	.3149	6.4	.6851	.3	.9909	20'
50'	.3119	6.1	.3212	6.3	.6788	.2	.9907	10'
12° 0'	9.3179	6.0	9.3275	6.3	10.6725	.3	9.9904	78° 0'
10'	.3238	5.9	.3336	6.1	.6664	.3	.9901	50'
20'	.3296	5.8	.3397	6.1	.6603	.2	.9899	40'
30'	.3353	5.7	.3458	6.1	.6542	.3	.9896	30'
40'	.3410	5.7	.3517	5.9	.6483	.3	.9893	20'
50'	.3466	5.6	.3576	5.9	.6424	.3	.9890	10'
13° 0'	9.3521	5.5	9.3634	5.8	10.6366	.3	9.9887	77° 0'
10'	.3575	5.4	.3691	5.7	.6309	.3	.9884	50'
20'	.3629	5.4	.3748	5.7	.6252	.3	.9881	40'
30'	.3682	5.3	.3804	5.6	.6196	.3	.9878	30'
40'	.3734	5.2	.3859	5.5	.6141	.3	.9875	20'
50'	.3786	5.2	.3914	5.5	.6086	.3	.9872	10'
14° 0'	9.3837	5.1	9.3968	5.4	10.6032	.3	9.9869	76° 0'
10'	.3887	5.0	.4021	5.3	.5979	.3	.9866	50'
20'	.3937	5.0	.4074	5.3	.5926	.4	.9863	40'
30'	.3986	4.9	.4127	5.3	.5873	.3	.9859	30'
40'	.4035	4.9	.4178	5.1	.5822	.3	.9856	20'
50'	.4083	4.8	.4230	5.2	.5770	.4	.9853	10'
15° 0'	9.4130	4.7	9.4281	5.1	10.5719	.3	9.9849	75° 0'
10'	.4177	4.7	.4331	5.0	.5669	.3	.9846	50'
20'	.4223	4.6	.4381	5.0	.5619	.4	.9843	40'
30'	.4269	4.6	.4430	4.9	.5570	.3	.9839	30'
40'	.4314	4.5	.4479	4.9	.5521	.4	.9836	20'
50'	.4359	4.5	.4527	4.8	.5473	.4	.9832	10'
16° 0'	9.4403	4.4	9.4575	4.8	10.5425	.3	9.9828	74° 0'
10'	.4447	4.4	.4622	4.7	.5378	.4	.9825	50'
20'	.4491	4.4	.4669	4.7	.5331	.4	.9821	40'
30'	.4533	4.2	.4716	4.7	.5284	.3	.9817	30'
40'	.4576	4.3	.4762	4.6	.5238	.4	.9814	20'
50'	.4618	4.2	.4808	4.6	.5192	.4	.9810	10'
17° 0'	9.4659	4.1	9.4853	4.5	10.5147	.4	9.9806	73° 0'
10'	.4700	4.1	.4898	4.5	.5102	.4	.9802	50'
20'	.4741	4.1	.4943	4.5	.5057	.4	.9798	40'
30'	.4781	4.0	.4987	4.4	.5013	.4	.9794	30'
40'	.4821	4.0	.5031	4.4	.4969	.4	.9790	20'
50'	.4861	4.0	.5075	4.4	.4925	.4	.9786	10'
18° 0'	9.4900	3.9	9.5118	4.3	10.4882	.4	9.9782	72° 0'
	L Cos	d 1'	L Cot	cd 1'	L Tan	d 1'	L Sin	Angle

III. LOGARITHMS OF TRIGONOMETRIC FUNCTIONS (*Cont.*)

Angle	L Sin	d 1'	L Tan	cd 1'	L Cot	d 1'	L Cos	
18° 0'	9.4900		9.5118		10.4882		9.9782	72° 0'
10'	.4939	3.9	.5161	4.3	.4839	.4	.9778	50'
20'	.4977	3.8	.5203	4.2	.4797	.4	.9774	40'
30'	.5015	3.8	.5245	4.2	.4755	.4	.9770	30'
40'	.5052	3.7	.5287	4.2	.4713	.5	.9765	20'
50'	.5090	3.8	.5329	4.2	.4671	.4	.9761	10'
19° 0'	9.5126	3.6	9.5370	4.1	10.4630	.4	9.9757	71° 0'
10'	.5163	3.7	.5411	4.1	.4589	.5	.9752	50'
20'	.5199	3.6	.5451	4.0	.4549	.4	.9748	40'
30'	.5235	3.6	.5491	4.0	.4509	.5	.9743	30'
40'	.5270	3.5	.5531	4.0	.4469	.4	.9739	20'
50'	.5306	3.6	.5571	4.0	.4429	.5	.9734	10'
20° 0'	9.5341	3.5	9.5611	4.0	10.4389	.4	9.9730	70° 0'
10'	.5375	3.4	.5650	3.9	.4350	.5	.9725	50'
20'	.5409	3.4	.5689	3.9	.4311	.4	.9721	40'
30'	.5443	3.4	.5727	3.8	.4273	.5	.9716	30'
40'	.5477	3.4	.5766	3.9	.4234	.5	.9711	20'
50'	.5510	3.3	.5804	3.8	.4196	.5	.9706	10'
21° 0'	9.5543	3.3	9.5842	3.8	10.4158	.4	9.9702	69° 0'
10'	.5576	3.3	.5879	3.7	.4121	.5	.9697	50'
20'	.5609	3.3	.5917	3.8	.4083	.5	.9692	40'
30'	.5641	3.2	.5954	3.7	.4046	.5	.9687	30'
40'	.5673	3.2	.5991	3.7	.4009	.5	.9682	20'
50'	.5704	3.1	.6028	3.7	.3972	.5	.9677	10'
22° 0'	9.5736	3.2	9.6064	3.6	10.3936	.5	9.9672	68° 0'
10'	.5767	3.1	.6100	3.6	.3900	.5	.9667	50'
20'	.5798	3.1	.6136	3.6	.3864	.6	.9661	40'
30'	.5828	3.0	.6172	3.6	.3828	.5	.9656	30'
40'	.5859	3.1	.6208	3.6	.3792	.5	.9651	20'
50'	.5889	3.0	.6243	3.5	.3757	.5	.9646	10'
23° 0'	9.5919	3.0	9.6279	3.6	10.3721	.6	9.9640	67° 0'
10'	.5948	2.9	.6314	3.5	.3686	.5	.9635	50'
20'	.5978	3.0	.6348	3.4	.3652	.6	.9629	40'
30'	.6007	2.9	.6383	3.5	.3617	.5	.9624	30'
40'	.6036	2.9	.6417	3.4	.3583	.6	.9618	20'
50'	.6065	2.9	.6452	3.5	.3548	.5	.9613	10'
24° 0'	9.6093	2.8	9.6486	3.4	10.3514	.6	9.9607	66° 0'
10'	.6121	2.8	.6520	3.4	.3480	.5	.9602	50'
20'	.6149	2.8	.6553	3.3	.3447	.6	.9596	40'
30'	.6177	2.8	.6587	3.4	.3413	.6	.9590	30'
40'	.6205	2.8	.6620	3.3	.3380	.6	.9584	20'
50'	.6232	2.7	.6654	3.4	.3346	.5	.9579	10'
25° 0'	9.6259	2.7	9.6687	3.3	10.3313	.6	9.9573	65° 0'
10'	.6286	2.7	.6720	3.3	.3280	.6	.9567	50'
20'	.6313	2.7	.6752	3.2	.3248	.6	.9561	40'
30'	.6340	2.7	.6785	3.3	.3215	.6	.9555	30'
40'	.6366	2.6	.6817	3.2	.3183	.6	.9549	20'
50'	.6392	2.6	.6850	3.3	.3150	.6	.9543	10'
26° 0'	9.6418	2.6	9.6882	3.2	10.3118	.6	9.9537	64° 0'
10'	.6444	2.6	.6914	3.2	.3086	.7	.9530	50'
20'	.6470	2.6	.6946	3.2	.3054	.6	.9524	40'
30'	.6495	2.5	.6977	3.1	.3023	.6	.9518	30'
40'	.6521	2.6	.7009	3.2	.2991	.6	.9512	20'
50'	.6546	2.5	.7040	3.1	.2960	.7	.9505	10'
27° 0'	9.6570	2.4	9.7072	3.2	10.2928	.6	9.9499	63° 0'
	L Cos	d 1'	L Cot	c d 1'	L Tan	d 1'	L Sin	Angle

III. LOGARITHMS OF TRIGONOMETRIC FUNCTIONS (Cont.)

Angle	L Sin	d 1'	L Tan	cd 1'	L Cot	d 1'	L Cos	
27° 0'	9.6570		9.7072		10.2928		9.9499	63° 0'
10'	.6595	2.5	.7103	3.1	.2897	.7	.9492	50'
20'	.6620	2.5	.7134	3.1	.2866	.6	.9486	40'
30'	.6644	2.4	.7165	3.1	.2835	.7	.9479	30'
40'	.6668	2.4	.7196	3.1	.2804	.6	.9473	20'
50'	.6692	2.4	.7226	3.0	.2774	.7	.9466	10'
28° 0'	9.6716	2.4	9.7257	3.1	10.2743	.7	9.9459	62° 0'
10'	.6740	2.4	.7287	3.0	.2713	.6	.9453	50'
20'	.6763	2.3	.7317	3.0	.2683	.7	.9446	40'
30'	.6787	2.4	.7348	3.1	.2652	.7	.9439	30'
40'	.6810	2.3	.7378	3.0	.2622	.7	.9432	20'
50'	.6833	2.3	.7408	3.0	.2592	.7	.9425	10'
29° 0'	9.6856	2.3	9.7438	3.0	10.2562	.7	9.9418	61° 0'
10'	.6878	2.2	.7467	2.9	.2533	.7	.9411	50'
20'	.6901	2.3	.7497	3.0	.2503	.7	.9404	40'
30'	.6923	2.2	.7526	2.9	.2474	.7	.9397	30'
40'	.6946	2.3	.7556	3.0	.2444	.7	.9390	20'
50'	.6968	2.2	.7585	2.9	.2415	.7	.9383	10'
30° 0'	9.6990	2.2	9.7614	2.9	10.2386	.8	9.9375	60° 0'
10'	.7012	2.2	.7644	3.0	.2356	.7	.9368	50'
20'	.7033	2.1	.7673	2.9	.2327	.7	.9361	40'
30'	.7055	2.2	.7701	2.8	.2299	.8	.9353	30'
40'	.7076	2.1	.7730	2.9	.2270	.7	.9346	20'
50'	.7097	2.1	.7759	2.9	.2241	.8	.9338	10'
31° 0'	9.7118	2.1	9.7788	2.9	10.2212	.7	9.9331	59° 0'
10'	.7139	2.1	.7816	2.8	.2184	.8	.9323	50'
20'	.7160	2.1	.7845	2.9	.2155	.8	.9315	40'
30'	.7181	2.1	.7873	2.8	.2127	.8	.9308	30'
40'	.7201	2.0	.7902	2.9	.2098	.7	.9300	20'
50'	.7222	2.1	.7930	2.8	.2070	.8	.9292	10'
32° 0'	9.7242	2.0	9.7958	2.8	10.2042	.8	9.9284	58° 0'
10'	.7262	2.0	.7986	2.8	.2014	.8	.9276	50'
20'	.7282	2.0	.8014	2.8	.1986	.8	.9268	40'
30'	.7302	2.0	.8042	2.8	.1958	.8	.9260	30'
40'	.7322	2.0	.8070	2.8	.1930	.8	.9252	20'
50'	.7342	2.0	.8097	2.7	.1903	.8	.9244	10'
33° 0'	9.7361	1.9	9.8125	2.8	10.1875	.8	9.9236	57° 0'
10'	.7380	1.9	.8153	2.8	.1847	.8	.9228	50'
20'	.7400	2.0	.8180	2.7	.1820	.9	.9219	40'
30'	.7419	1.9	.8208	2.8	.1792	.8	.9211	30'
40'	.7438	1.9	.8235	2.7	.1765	.8	.9203	20'
50'	.7457	1.9	.8263	2.8	.1737	.9	.9194	10'
34° 0'	9.7476	1.9	9.8290	2.7	10.1710	.8	9.9186	56° 0'
10'	.7494	1.8	.8317	2.7	.1683	.9	.9177	50'
20'	.7513	1.9	.8344	2.7	.1656	.8	.9169	40'
30'	.7531	1.8	.8371	2.7	.1629	.9	.9160	30'
40'	.7550	1.9	.8398	2.7	.1602	.9	.9151	20'
50'	.7568	1.8	.8425	2.7	.1575	.9	.9142	10'
35° 0'	9.7586	1.8	9.8452	2.7	10.1548	.8	9.9134	55° 0'
10'	.7604	1.8	.8479	2.7	.1521	.9	.9125	50'
20'	.7622	1.8	.8506	2.7	.1494	.9	.9116	40'
30'	.7640	1.8	.8533	2.6	.1467	.9	.9107	30'
40'	.7657	1.7	.8559	2.7	.1441	.9	.9098	20'
50'	.7675	1.8	.8586	2.7	.1414	.9	.9089	10'
36° 0'	9.7692	1.7	9.8613		10.1387	.9	9.9080	54° 0'
	L Cos	d 1'	L Cot	cd 1'	L Tan	d 1'	L Sin	Angle

III. LOGARITHMS OF TRIGONOMETRIC FUNCTIONS (Cont.)

Angle	L Sin	d 1'	L Tan	cd 1'	L Cot	d 1'	L Cos	
36° 0'	9.7692		9.8613		10.1387		9.9080	54° 0'
10'	.7710	1.8	.8639	2.6	.1361	1.0	.9070	50'
20'	.7727	1.7	.8666	2.7	.1334	.9	.9061	40'
30'	.7744	1.7	.8692	2.6	.1308	.9	.9052	30'
40'	.7761	1.7	.8718	2.6	.1282	1.0	.9042	20'
50'	.7778	1.7	.8745	2.7	.1255	.9	.9033	10'
37° 0'	9.7795	1.7	9.8771	2.6	10.1229	1.0	9.9023	53° 0'
10'	.7811	1.6	.8797	2.6	.1203	.9	.9014	50'
20'	.7828	1.7	.8824	2.7	.1176	1.0	.9004	40'
30'	.7844	1.6	.8850	2.6	.1150	.9	.8995	30'
40'	.7861	1.7	.8876	2.6	.1124	1.0	.8985	20'
50'	.7877	1.6	.8902	2.6	.1098	1.0	.8975	10'
38° 0'	9.7893	1.6	9.8928	2.6	10.1072	1.0	9.8965	52° 0'
10'	.7910	1.7	.8954	2.6	.1046	1.0	.8955	50'
20'	.7926	1.6	.8980	2.6	.1020	1.0	.8945	40'
30'	.7941	1.5	.9006	2.6	.0994	1.0	.8935	30'
40'	.7957	1.6	.9032	2.6	.0968	1.0	.8925	20'
50'	.7973	1.6	.9058	2.6	.0942	1.0	.8915	10'
39° 0'	9.7989	1.6	9.9084	2.6	10.0916	1.0	9.8905	51° 0'
10'	.8004	1.5	.9110	2.6	.0890	1.0	.8895	50'
20'	.8020	1.6	.9135	2.5	.0865	1.1	.8884	40'
30'	.8035	1.5	.9161	2.6	.0839	1.0	.8874	30'
40'	.8050	1.5	.9187	2.6	.0813	1.0	.8864	20'
50'	.8066	1.6	.9212	2.5	.0788	1.1	.8853	10'
40° 0'	9.8081	1.5	9.9238	2.6	10.0762	1.0	9.8843	50° 0'
10'	.8096	1.5	.9264	2.6	.0736	1.1	.8832	50'
20'	.8111	1.5	.9289	2.5	.0711	1.1	.8821	40'
30'	.8125	1.4	.9315	2.6	.0685	1.1	.8810	30'
40'	.8140	1.5	.9341	2.6	.0659	1.0	.8800	20'
50'	.8155	1.5	.9366	2.5	.0634	1.1	.8789	10'
41° 0'	9.8169	1.4	9.9392	2.6	10.0608	1.1	9.8778	49° 0'
10'	.8184	1.5	.9417	2.5	.0583	1.1	.8767	50'
20'	.8198	1.4	.9443	2.6	.0557	1.1	.8756	40'
30'	.8213	1.5	.9468	2.5	.0532	1.1	.8745	30'
40'	.8227	1.4	.9494	2.6	.0506	1.2	.8733	20'
50'	.8241	1.4	.9519	2.5	.0481	1.1	.8722	10'
42° 0'	9.8255	1.4	9.9544	2.5	10.0456	1.1	9.8711	48° 0'
10'	.8269	1.4	.9570	2.6	.0430	1.2	.8699	50'
20'	.8283	1.4	.9595	2.5	.0405	1.1	.8688	40'
30'	.8297	1.4	.9621	2.6	.0379	1.2	.8676	30'
40'	.8311	1.4	.9646	2.5	.0354	1.1	.8665	20'
50'	.8324	1.3	.9671	2.5	.0329	1.2	.8653	10'
43° 0'	9.8338	1.4	9.9697	2.6	10.0303	1.2	9.8641	47° 0'
10'	.8351	1.3	.9722	2.5	.0278	1.2	.8629	50'
20'	.8365	1.4	.9747	2.5	.0253	1.1	.8618	40'
30'	.8378	1.3	.9772	2.5	.0228	1.2	.8606	30'
40'	.8391	1.3	.9798	2.6	.0202	1.2	.8594	20'
50'	.8405	1.4	.9823	2.5	.0177	1.2	.8582	10'
44° 0'	9.8418	1.3	9.9848	2.5	10.0152	1.3	9.8569	46° 0'
10'	.8431	1.3	.9874	2.6	.0126	1.2	.8557	50'
20'	.8444	1.3	.9899	2.5	.0101	1.2	.8545	40'
30'	.8457	1.3	.9924	2.5	.0076	1.3	.8532	30'
40'	.8469	1.2	.9949	2.5	.0051	1.2	.8520	20'
50'	.8482	1.3	.9975	2.6	.0025	1.3	.8507	10'
45° 0'	9.8495	1.3	10.0000	2.5	10.0000	1.2	9.8495	45° 0'
	L Cos	d 1'	L Cot	cd 1'	L Tan	d 1'	L Sin	Angle

IV. CONSTANTS AND THEIR LOGARITHMS

Constant	Value	Common Logarithm
π	3.1415 9265	0.4971 4987
e	2.7182 8183	0.4342 9448
$\mu = \log_{10} e$	.4342 9448	9.6377 8431 − 10
$\dfrac{1}{\mu} = \log_e 10$	2.3025 8509	0.3622 1569
1 radian = $\dfrac{180}{\pi}$ degrees	57°.2957 7951 = 57° 17′ 44″.8	1.7581 2263
1 degree = $\dfrac{\pi}{180}$ radians	0.0174 5329	8.2418 7737 − 10
1 minute = $\dfrac{\pi}{10,800}$ radians	0.0002 9089	6.4637 2612 − 10
1 second = $\dfrac{\pi}{648,000}$ radians	0.0000 0485	4.6855 7487 − 10
$\sqrt{\pi}$	1.7724 5385	0.2485 7494
$\dfrac{1}{\pi}$	.3183 0989	9.5028 5013 − 10
$\dfrac{1}{\sqrt{\pi}}$	0.5641 8958	9.7514 2506 − 10
$\dfrac{1}{\sqrt[3]{\pi}}$	0.6827 8406	9.8342 8338 − 10
$\sqrt{e}$	1.6487 2107	.2171 4724
$\dfrac{1}{e}$	0.3678 7944	9.5657 0552 − 10
g	$32.16 \dfrac{\text{ft.}}{\text{sec.}^2} = 981 \dfrac{\text{cm.}}{\text{sec.}^2}$	
centimeters in 1 foot	30.480	1.4840 158
feet in 1 centimeter	0.0328 08	8.5159 842 − 10
inches in 1 meter	39.37	1.5951 654
kilometers in 1 mile	1.6093	.20664
kilograms in 1 pound	0.4536	9.6566 660 − 10
pounds in 1 kilogram	2.2046	.3433 340
weight 1 cubic foot water	62.425	1.7953 586
cubic inches in 1 gallon	231	2.3636 120

INDEX

(The numbers refer to pages.)

The student should note that the index forms an excellent basis for reviewing the course. Consider each topic, recalling to mind the details concerning each one. Whenever in doubt, check by reading the reference.

Abscissa, 23
Acute angles, 8
Ambiguous case, 63
Angle, complementary, 12
  cosecant of an, 7
  cosine of an, 7
  cotangent of an, 7
  coterminal, 4, 36, 77
  defined, 3
  negative, 3
  of depression, 21
  of elevation, 21
  positive, 3
  quadrantal, 35
  secant of an, 7
  sine of an, 7
  tangent of an, 7
  trisection of an, 43
Angular measurement, 4
Answers, 95
Application of logarithms, 55
Arc, length of, 5
Arcsin, 20
Area of a triangle, 2
Areas of triangles, 71

Base of logarithms, 48

Characteristic, 50
  rule for, 52
Circle, circumference of, 2
  radius of circumscribed, 93
  radius of inscribed, 72
  segment of, 89
  unit radius, 28

Co-function, 11
Common logarithms, 48
Complementary angles, 12
Complementary relationships, 11
Computer's rule, 53
Coordinates, 23
  polar, 24
  rectangular, 23
  systems of, 23
Cosecant of an angle, 7
Cosine, graph of, 34
  of an angle, 7
  of a difference, 39
  of a sum, 38
Cosines, law of, 67
Cotangent of an angle, 7
Coterminal angles, 4, 36, 77

Decreasing function, 19
Definitions, extension of, 23
Degrees to radians, 5
Depression, angle of, 21
Difference, cosine of a, 39
  sine of a, 39
  tangent of a, 40
Direct functions, 36
Directions for establishing identities, 16
Double angles, functions of, 42

Elevation, angle of, 21
Equations, trigonometric, 37
Exercises, miscellaneous, 89
  supplementary, 77
Exponential form, 48

INDEX

Form, exponential, 48
  logarithmic, 48
Formulas, product, 46
  reduction, 30
Functions, decreasing, 19
  direct, 36
  increasing, 19
  inverse, 36
  of double angles, 42
  of half angles, 43
  signs of, 25
  trigonometric, 7
  45°, 30°, 60°, 10
Fundamental identities, 13

Graph, of cosine, 34
  of sine, 33
  of tangent, 34

Half angles, functions of, 43

Identities, 13
  fundamental, 15
Increasing functions, 19
Infinity, 35
Initial side, 3
Inscribed circle, radius of, 72
Interpolation, 18, 53
Inverse functions, 20, 36
Inverse sine, 20

Law, of cosines, 67
  of sines, 60
  of tangents, 68
Laws of logarithms, 49
Length of arc, 5
Line values of functions, 28
Logarithm, of a negative number, 52
  of a power, 49
  of a product, 49
  of a quotient, 49
  of a root, 49
Logarithmic form, 48
Logarithms, 48
  applications, 55
  common, 48
  laws of, 49

Mantissa, 50
Miscellaneous exercises, 89

Negative angle, 3
Negative number, logarithm of, 52
Negative radius vector, 23

Oblique triangles, 60
Ordinate, 23
Origin, 23

Period of the functions, 36
Periodicity, 35
Polar coordinates, 24
Position, standard, 24
Positive angle, 3
Power, logarithm of a, 49
Problems, miscellaneous, 89
  supplementary, 77
Product, logarithm of a, 49
Product formulas, 46
Projection, backward, 29
Proportion, defined, 1
Proportional parts, principle of, 53
Pythagorean theorem, 1

Quadrantal angles, 35
Quadrants, 23
Quotient, logarithm of a, 49

Radian measure, 4
Radians to degrees, 5
Radius, of circumscribed circle, 93
  of inscribed circle, 72
Radius vector, 23
Ratio, defined, 1
Reciprocal pairs, 13
Reciprocal relationships, 13
Rectangular coordinates, 23
Reduction formulas, 30
Reference, triangle of, 24
Right triangles, 18
  solution of, 19
  solution using logarithms, 57
Root, logarithm of a, 49
Rule, for characteristic, 52
  for signs of functions, 32

INDEX

Secant of an angle, 7
Segment of a circle, 89
Sexagesimal system, 4
Side, initial, 3
   terminal, 3
Signs, of functions, 25
   rule for, 32
Similar triangles, 2
Sine, graph of, 33
   inverse, 20
   of an angle, 7
   of a difference, 39
   of a sum, 38
Sines, law of, 60
Solution of triangles, 19, 57
   case 1, 61
   case 2, 63
   case 3, 69
   case 4, 74
Standard position of reference angle, 24
Sum, cosine of, 38
   sine of a, 38
   tangent of a, 39
Supplementary exercises, 77

Tables, constants and their logarithms, 118
   functions of angles, 108
   logarithms of functions, 113
   logarithms of numbers, 106
   use of, 18, 50

Tangent, graph of, 34
Tangent, of a difference, 40
   of a sum, 39
   of an angle, 7
Tangents, law of, 68
Terminal side, 2
Theorem, Pythagorean, 1
Triangle, area of a, 2
   areas of (oblique), 71
Triangle, solving the, 19
   case 1, 61
   case 2, 63
   case 3, 69
   case 4, 74
Triangle of reference, 24
Triangles, right, 18, 57
   similar, 2
   solution of, 19, 57, 60
Trigonometric, equations, 37
   functions, 7
Trisection of an angle, 43

Unit circle, 28
Units of angular measurement, 4
Use of tables, 18, 50

Values, line, 28
Variation of the functions, 35
Vector, radius, 23

Zero angle, functions of, 35

www.ingramcontent.com/pod-product-compliance
Lightning Source LLC
Chambersburg PA
CBHW080251170426
43192CB00014BA/2633